乳中活性因子与人体健康系列丛书

乳中活性因子与人体健康

——短链脂肪酸

◎ 孟 璐 郑 楠 主编

中国农业科学技术出版社

图书在版编目（CIP）数据

乳中活性因子与人体健康．短链脂肪酸 / 孟璐，郑楠主编．-- 北京：中国农业科学技术出版社，2023.11
ISBN 978-7-5116-6487-7

Ⅰ．①乳…　Ⅱ．①孟…②郑…　Ⅲ．①乳液—脂肪酸—关系—健康—研究　Ⅳ．① Q592.6 ② R16

中国国家版本馆 CIP 数据核字（2023）第 203086 号

责任编辑　金　迪
责任校对　李向荣
责任印制　姜义伟　王思文

出 版 者　中国农业科学技术出版社
北京市中关村南大街 12 号　邮编：100081
电　　话　（010）82106625（编辑室）（010）82106624（发行部）
（010）82109709（读者服务部）
网　　址　https://castp.caas.cn
经 销 者　各地新华书店
印 刷 者　北京建宏印刷有限公司
开　　本　170 mm × 240 mm　1/16
印　　张　4.75　　折页 1
字　　数　73 千字
版　　次　2023 年 11 月第 1 版　2023 年 11 月第 1 次印刷
定　　价　36.00 元

乳中活性因子与人体健康——短链脂肪酸

编 委 会

主　编　孟　璐　郑　楠

副主编　张养东　刘慧敏　赵圣国　高亚男

参　编　（以姓氏拼音为序）

黄胥莱　席雪瑶　杨亚新　姚倩倩

序　言

奶，常被称为“自然界最接近完美的食物”，不仅在传统上被视为营养的源泉，更因其丰富的活性成分而备受关注。在当今大健康时代，我们对奶的理解已超越了对其基本营养功能的认识，深入到了它所包含的各种活性营养成分及其对人体健康的深远影响。

奶中含有多种活性因子，这些成分不仅提供基础营养，还具有重要的生物功能。例如抗菌、抗炎、抗病毒和免疫调节功能，还有促进细胞生长和修复，促进大脑、骨骼、肠道等器官发育等作用。

近年来，学者们对奶中的活性因子进行了大量研究，这些研究揭示了奶中活性因子的多种健康益处和潜在机制。奶中的活性营养成分共同作用，为人体提供了全面、均衡的营养，同时在促进健康、预防疾病方面展现了多种生物功能。深入研究奶中的活性营养成分，不仅有助于我们更好地理解奶的营养价值，还能为奶制品的开发和应用提供科学依据，最终为人类健康带来更多福祉。

在《乳中活性因子与人体健康系列丛书》中，我们将系统探讨奶中的主要活性营养成分及其生物功能，揭示这“自然界最接近完美的食物”内在的科学奥秘，期望为读者提供全新的视角和深入的理解。旨在帮助读者掌握科学的饮食知识，合理地将奶制品融入日常饮食中。通过这套书，读者不仅可以提升对奶制品的认识，还能学会如何在生活中更好地利用这些营养成分，以提高自己的生活质量和健康水平。我们真诚地希望，这套书能够成为广大读者的健康指南，带给大家更加美好和健康的生活。

前　言

短链脂肪酸是一组碳原子数小于 6 的有机酸，反刍动物乳中的短链脂肪酸含量为 1% ～ 4%，人乳中的短链脂肪酸含量通常较低，主要是由肠道微生物代谢过程产生。近年来随着研究的深入，人们发现其在婴幼儿的生长和发育过程中扮演着关键角色，尤其在维护肠道健康和免疫系统发育方面。短链脂肪酸主要由肠道微生物发酵碳水化合物而来，具有抗炎、促进肠道发育、保护肾脏、肺等多脏器健康以及促进大脑发育、提高认知功能重要生理功能。此外短链脂肪酸还可作为能量来源，支持婴幼儿的生长和发育。短链脂肪酸的广泛生物学功能决定了其具有多重应用场景，如婴幼儿配方奶粉、功能性食品、食品工业等方面。

针对短链脂肪酸的高生物活性和多重生理功能特点，本书重点介绍脂肪酸的结构、检测方法、生物学作用以及应用，旨在为短链脂肪酸活性功能探究及脂肪酸富集乳产品产业化，特别是婴幼儿乳品，提出新思路。

由于作者水平有限，书籍中疏漏之处在所难免，敬请读者批评指正。

目 录

1 短链脂肪酸的生物学特性

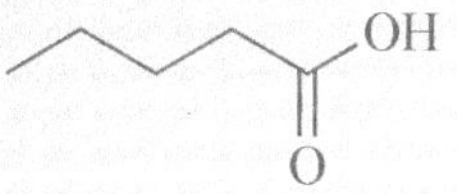

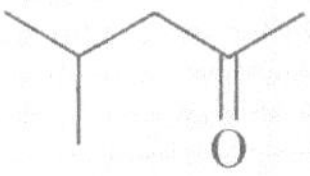

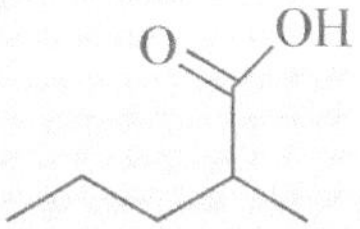

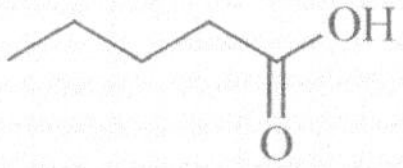

1.1 乳短链脂肪酸的含量

1.1.1 短链脂肪酸定义及理化性质

短链脂肪酸（SCFAs）是一组碳原子数小于 6 的有机酸（Annunziata 等，2020；Cai 等，2017），C1:0（甲酸、蚁酸，CAS 号：64–18–6，分子量：46）、C2:0（乙酸、醋酸、冰醋酸，CAS 号：64–19–7，分子量：66）、C3:0（丙酸、初油酸，CAS 号：79–09–4，分子量：74）、*iso* C4:0（异丁酸、2- 甲基丙酸，CAS 号：79–31–2，分子量：88）、C4:0（正丁酸、酪酸，CAS 号：107–92–6，分子量：88）、*iso* C5:0（2- 甲基丁酸，CAS 号：116–53–0，分子量：102）、*anteiso* C5:0（异戊酸、3- 甲基丁酸，CAS 号：503–74–2，分子量：102）、C5:0（正戊酸，CAS 号：109–52–4，分子量：102）、*iso* C6:0（异己酸、2- 甲基戊酸、二氢草莓酸，CAS 号：97–61–0，分子量：116）、C6:0（正己酸、羊油酸，CAS 号：142–62–1，分子量：116）10 种 SCFAs 化学结构见图 1.1。

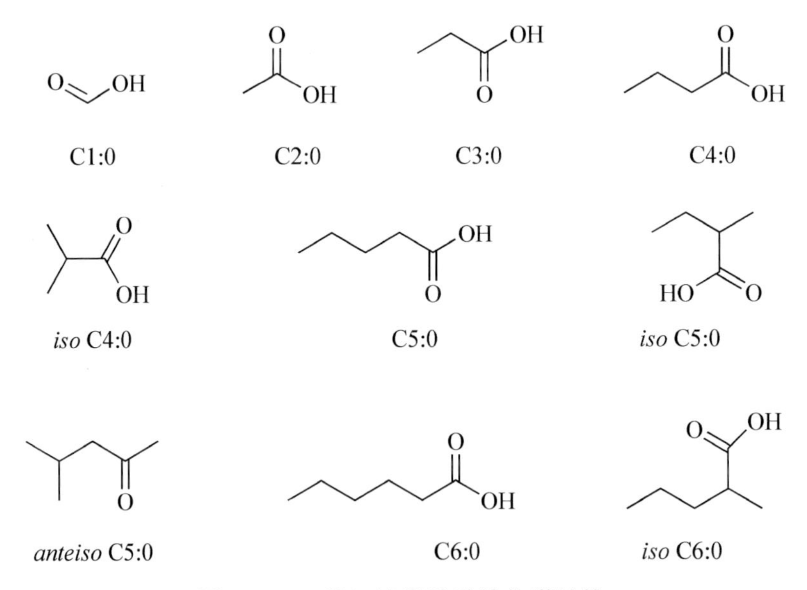

图 1.1　10 种短链脂肪酸的化学结构

SCFAs 理化性质如表 1.1 所示，SCFAs 为无色或淡黄色透明液体，大部分具有刺激性的气味。SCFAs 密度与水相近，易挥发，沸点随着碳链的延长而增加。SCFAs 溶于水，也溶于乙醇、乙醚等试剂。

表 1.1　10 种短链脂肪酸理化性质

种类	外观性状	密度（g/cm³）	沸点（℃）	溶解性
C1:0	无色透明液体，强烈刺激性气味	1.22	100.6	易溶于水，溶于乙醇、乙醚、氯仿
C2:0	无色透明液体，强烈刺激性醋酸味	1.05	117.9	易溶于水，溶于乙醇、乙醚、氯仿
C3:0	无色透明液体，强烈刺激性气味	0.99	141.1	易溶于水，溶于乙醇、乙醚、氯仿
iso C4:0	浅色透明液体，强烈刺激性气味	0.95	153.5	易溶于水，溶于乙醇、乙醚
C4:0	无色油状液体，刺激性、难闻气味	0.96	164.4	易溶于水、溶于乙醇、乙醚
iso C5:0	无色至淡黄色透明液体，羊乳干酪气味	0.94	175.3	溶于水、乙醇、乙醚
anteiso C5:0	无色至淡黄色透明液体，微有臭味	0.93	176.5	溶于水、乙醇、乙醚
C5:0	无色至淡黄色透明液体，刺激性气味	0.94	185.3	溶于水、乙醇、乙醚
iso C6:0	无色至淡黄色液体，刺激性辣味	0.92	195.5	微溶于水，溶于乙醇、乙醚
C6:0	无色或淡黄色油状液体，汗臭味	0.93	205.4	微溶于水，溶于乙醇、乙醚

1.1.2　不同动物乳中短链脂肪酸含量

乳中 SCFAs 以结合态和游离态两种形式存在，结合态 SCFAs 主要以酯化形式结合于甘油三酯的 sn–3 和 sn–1 位，游离态则以阴离子的形式存在（Dai 等，2020；Stumpff，2018）。乳中 SCFAs 含量较低，文献报道中乳 SCFAs 的结果表现形式包括绝对含量和相对含量两种。

前人对乳中 SCFAs 绝对含量的研究较少，且以往研究乳中 SCFAs 的检测结果较为碎片化，检测数据结果形式并不一致。在统一度量单位标准后，表 1.2 总结了近年来文献中生乳 SCFAs 的绝对含量（μg/mL）。正如表 1.2 所示，与人乳 SCFAs 相比，研究牛乳 SCFAs 绝对含量的文献较少，水牛乳、山羊乳、绵羊乳、牦牛乳、骆驼乳、驴乳、马乳等特色乳 SCFAs 绝对含量几乎未被研究。研究发现，牛乳中 C1:0 和 *iso* C6:0、人乳中的 *iso* C5:0、*anteiso* C5:0、C5:0、*iso* C6:0 均未被检测及研究，人乳中 C1:0 和 C2:0 的含量显著高于牛乳，C4:0 和 C6:0 的含量显著低于牛乳，人乳和牛乳中 C3:0、*iso* C4:0 含量均较低或未检出。

表 1.2　不同哺乳动物乳中常见短链脂肪酸绝对含量（单位 :μg/mL）

种类	来源					
	奶牛			人		
C1:0	ND	ND	ND	4.98	174.91	22.68
C2:0	ND	ND	1.51	11.71	258.00	40.32
C3:0	ND	ND	0.12	0.48	ND	ND
iso C4:0	ND	ND	0.05	0.07	ND	ND
C4:0	335.00	1468.00	8.33	11.26	17.68	23.28
iso C5:0	ND	ND	0.02	ND	ND	ND
anteiso C5:0	ND	ND	0.03	ND	ND	ND
C5:0	ND	ND	0.12	ND	ND	ND
*iso*C6:0	ND	ND	ND	ND	ND	ND
C6:0	501.00	509.00	3.65	15.82	ND	ND
参考文献	Liu 等，2018a	Liu 等，2020	Li 等，2022	Jiang 等，2016	Prentice 等，2019	Stinson 等，2020

注：ND，代表未检出。

目前仅有对乳中 C4:0 和 C6:0 相对含量的研究报道，而对于其他 SCFAs 含量鲜有涉及。表 1.3 总结了近年来文献报道的生乳中 C4:0 和 C6:0 在总脂肪酸中的相对含量（g/100g 脂肪酸）。结果显示，反刍动物（奶牛、水牛、山羊、绵羊、牦牛等）乳中的 SCFAs 含量为 1% ~ 4%（Chamekh 等，2020；Liu 等，2011；Pegolo 等，2017；Teng 等，2017；Yurchenko 等，2018），单胃动物（人、驴、马等）乳中 SCFAs 的含量少于1%（Dai等，2020；Martini等，2015；Wan等，2010）。然而，Wang等（2022a）和Chamekh等（2020）均发现骆驼乳中SCFAs的含量少于1%，与其他反刍动物（奶牛、水牛、山羊、绵羊、牦牛等）乳差异较大，而与单胃动物（人、驴、马等）乳较为相似。

表1.3 不同哺乳动物乳中常见短链脂肪酸相对含量
（单位：g/100g脂肪酸）

来源	C4:0	C6:0	参考文献
奶牛	2.52 ± 0.51	1.50 ± 0.26	Wang等，2022a
	3.14 ± 0.27	2.17 ± 0.25	Pietrzak-Fiécko等，2020
	3.75 ± 0.04	2.28 ± 0.07	Khan等，2017
	3.45 ± 0.90	2.15 ± 0.38	Pegolo等，2017
	3.77	2.32	Gastaldi等，2010
	1.23 ± 0.02	1.75 ± 0.03	Teng等，2017
水牛	2.96 ± 0.81	1.20 ± 0.32	Wang等，2022a
	3.98 ± 0.09	2.41 ± 0.13	Khan等，2017
	2.58 ~ 5.41	1.41 ~ 4.04	Nie等，2022
	3.80 ± 0.57	2.31 ± 0.45	Pegolo等，2017
	1.62 ± 0.03	2.09 ± 0.04	Teng等，2017
牦牛	2.12 ± 0.13	1.49 ± 0.20	Wang等，2022a
	2.05 ± 0.03	3.42 ± 0.05	Teng等，2017
	3.07 ± 0.36	2.53 ± 0.52	Liu等，2011
山羊	1.71 ± 0.33	1.57 ± 0.38	Wang等，2022a
	2.56 ± 0.12	2.79 ± 0.04	Pietrzak-Fiécko等，2020
	1.05 ± 0.04	1.77 ± 0.16	Korma等，2022
	2.09 ± 0.04	3.64 ± 0.08	Yurchenko等，2018
绵羊	2.81 ± 0.14	2.54 ± 0.13	Pietrzak-Fiécko等，2020
	1.81 ± 0.28	1.52 ± 0.33	Ptacek等，2019
骆驼	0.02 ± 0.01	0.07 ± 0.02	Wang等，2022a
	0.05 ± 0.01	0.14 ± 0.01	Chamekh等，2020

续表

来源	C4:0	C6:0	参考文献
马	0.18 ± 0.09	0.28 ± 0.26	Pietrzak–Fiécko 等，2020
	0.19	0.20	Czyzak–Runowska 等，2021
驴	0.29 ± 0.13	0.07 ± 0.02	Wang 等，2022a
	0.04 ~ 0.15	0.11 ~ 0.14	Messias 等，2021
	0.57	0.16	Gastaldi 等，2010
	0.03 ~ 0.05	0.20 ~ 0.27	Martini 等，2015
人	0.03 ± 0.02	0.03 ± 0.02	Wang 等，2022a
	0.02 ± 0.03	0.09 ± 0.05	Pietrzak–Fiécko 等，2020
	0.01	0.02	Gastaldi 等，2010
	0.10 ± 0.01	0.11 ± 0.02	Teng 等，2017
	0.06 ± 0.01	0.07 ± 0.02	Wan 等，2010

1.2 乳短链脂肪酸的合成

乳 SCFAs 是瘤胃/肠道微生物群发酵副产物通过循环到乳腺或在乳腺内合成（Koh 等，2016；Stinson 等，2020）。直链 SCFAs，如 C1:0、C2:0、C3:0、C4:0、C5:0 和 C6:0 等主要来源于瘤胃/肠道厌氧细菌或酵母菌发酵未被消化吸收的碳水化合物，如抗性淀粉、饲料纤维等。研究发现，C1:0 主要产生菌为乳酸杆菌和具核梭杆菌；C2:0 主要产生菌为双歧杆菌属、乳酸杆菌属、嗜黏蛋白阿尔曼氏菌、普氏菌属、瘤胃球菌属和拟杆菌门；C3:0 主要产生菌为拟杆菌门、厚壁菌门、肠杆菌科和毛螺菌科；C4:0 主要产生菌为厚壁菌门、普氏栖粪杆菌、直肠真杆菌和罗斯氏菌属；C5:0 主要产生菌为拟杆菌门、普雷沃氏菌科和厚壁菌门；C6:0 主要产生菌为厚壁菌门韦荣球菌科和瘤胃菌科（Agus 等，2021；Koh 等，2016；Smith 等，2013）。当饲料能量供应不足时，瘤胃/肠道微生物发酵碳水化合物转向氨基酸、内源蛋白质、饲料脂肪等，如发酵缬氨酸、亮氨酸、异亮氨酸产生支链 SCFAs，如 *iso* C4:0、*iso* C5:0、*anteiso* C5:0 和 *iso* C6:0（Koh

等，2016；Prentice 等，2019）。瘤胃 / 肠道内的 SCFAs 绝大部分以 $SCFA^-$ 的形式存在，其余以未解离 SCFAs 的形式存在（Stumpff，2018），通过主动转运（$SCFA^-$–HCO_3^- 交换、单羧酸转运蛋白、钠离子偶联单羧酸转运蛋白 1 等）和被动扩散吸收进入血液中（Sivaprakasam 等，2017）。血液中的脂类被乳腺泡毛细血管内脂蛋白酶水解后经基底细胞膜运输至乳腺组织，进一步被重新合成结合态甘油三酯。乳腺摄取血液中游离态 SCFAs 的能力有限，反刍动物血液中的 C2:0、β – 羟基丁酸等经血液运输至乳腺之后作为前体物质参与乳中 SCFAs 的从头合成，单胃动物则利用血液中的葡萄糖作为乳脂 SCFAs 合成的前体物（Fox 等，2015）。SCFAs 的合成如图 1.2 所示。

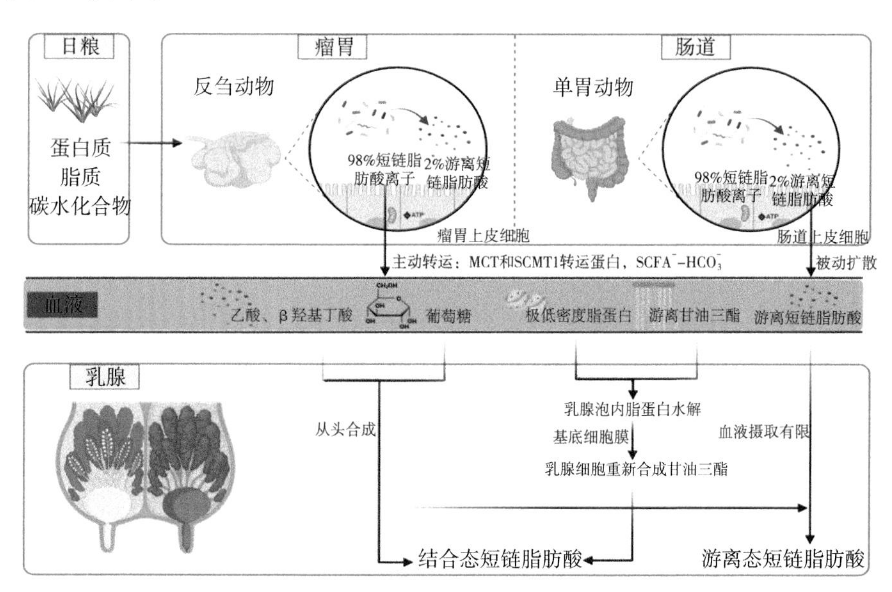

图 1.2　生乳短链脂肪酸合成示意图（武旭芳等，2022）

1.2.1　植物性食物来源

（1）以糖为底物合成 SCFAs

木聚糖酶、葡聚糖酶和阿拉伯糖酶等膳食纤维是肠道菌群产生 SCFAs 的主要底物。糖苷水解酶及其肠道菌群将膳食纤维降解为单糖，主要是戊糖和已糖（Wang 等，2019），然后通过糖酵解或戊糖磷酸途径生成 SCFAs（Koh 等，2016）。乙酸主要由丙酮酸通过两步酶催化（丙酮酸甲酸裂解酶和乙酸-辅酶A连接酶）与代谢中间体乙酰-辅酶A产生。产酸细菌，例如梭菌属的某些菌种，可以将1个葡萄糖分子转化为3个乙酸分子（Karnholz等，2002）。丙酸合成最重要的途径之一是琥珀酸途径，拟杆菌门如*Dialister* spp.和*Veillonella* sp.等以戊糖或己糖为底物合成丙酸。此外，毛螺菌科和瘤胃球菌科可以将岩藻糖或鼠李糖等其他糖类作为丙二醇途径的底物合成丙酸（Reichardt等，2020）。

（2）通过碳链延伸途径合成SCFAs

通过辅酶A依赖性途径的碳链延长是丁酸合成的经典途径：2分子乙酰辅酶A可缩合生成1分子丁酰辅酶A，然后通过丁酰辅酶A和乙酸辅酶A转移酶转化成丁酸（Duncan等，2002）。丁酰辅酶A也可以通过磷酸转丁酰酶和丁酸激酶转化为丁酸，但磷酸转丁酰酶和丁酸激酶在人结肠中不常见（Louis等，2014）。之前的研究已经分离出了产生戊酸的*Megasphaera elsdenii*和*Megasphaera* sp.，戊酸合成的关键步骤由硫解酶催化，丙酰辅酶A和乙酰辅酶A作为底物（Yoshikawa等，2018），但鲜有对戊酸合成的代谢途径进行进一步研究。目前还没有确定的己酸合成途径，但通常认为丁酰辅酶A和乙酰辅酶A通过硫解酶缩合合成己酸（Xu等，2020）。

（3）其他代谢途径

除了上述提到的两种合成途径外，SCFAs还可以由Wood-Ljungdahl途径合成。某些肠道细菌如*Blautia hydrogenotrophica*可以使用H_2和CO_2作为底物来合成乙酸，该菌株还可以将甲酸作为底物通过Wood-Ljungdahl途径合成乙酸（Yoshikawa等，2018）。此外，一些梭菌属通过丙烯酸酯途径，以乳酸作为前体物质通过丙烯酰辅酶A合成丙酸，但这种方式产生的丙酸有限（Flint等，2015）。丁酸也可以使用乳酸和乙酸作为底物，由厌氧菌属等和真杆菌属合成（Rey等，2010）。乳酸是SCFAs重要的代谢中间产物，可作为*Megasphaera elsdenii*合成戊酸的底物（Louis等，2014），或被梭菌属转化为丁酸或己酸（Zhu等，2015）。

1.2.2 动物性食物来源

以氨基酸为底物合成支链SCFAs：通过氨基酸代谢途径与α-酮酸的脱羧和还原途径相结合，可产生支链SCFAs，如异丁酸、异戊酸和异己酸（Atsumi 等，2008）。大约 17% 和 38% 的支链 SCFAs 分别通过盲肠和乙状结肠以及直肠中的蛋白质发酵产生（Macfarlane 等，1992）。缬氨酸、亮氨酸和异亮氨酸等氨基酸可以作为合成异丁酸、异戊酸、异己酸和 2-甲基丁酸的底物。目前有研究已经分离出能够合成支链戊酸的天然肠源性细菌 *Megasphaera elsdenii*、*Prevotella copri* 和 *Prevotella stercorea*，为肠道菌群合成支链脂肪酸提供了强有力的支撑（Louis 等，2014）。此外，氨基酸的降解也会影响乙酸和丙酸的合成，如谷氨酸、丙氨酸、天冬氨酸代谢会影响三羧酸循环和丙酸前体琥珀酸的代谢。

1.3 短链脂肪酸的吸收途径

SCFAs 优先从三酰甘油分子中水解，并直接从肠道转移到血液中。大多数 SCFAs 在肠道附近产生和利用，其中一小部分丙酸和乙酸到达肝脏，通过线粒体 β - 氧化快速代谢，并可用作产生能量的三羧酸循环的底物，有效代谢产生葡萄糖，成为机体的快速能量来源。

肠道中只有一小部分联合形式存在的 SCFAs 可直接穿过上皮屏障，而绝大部分 SCFAs 以电离状态存在，需要专门的转运蛋白才能被机体吸收。大部分 SCFAs 主要通过单羧酸转运蛋白 1（monocarboxylate transporter 1，MCT-1）和钠偶联单羧酸转运蛋白 1（sodium-dependent monocarboxylate transporter 1，SMCT-1）这两种受体蛋白介导的主动运输穿过黏膜通道。MCT-1 和 SMCT-1 在结肠细胞以及包括小肠和盲肠在内的整个胃肠道上都高度表达（Iwanaga 等，2006）。此外，MCT-1 在淋巴细胞上也高度表达，SMCT-1 在肾脏和甲状腺上表达，SMCT-1 结合 SCFAs 的亲和力顺序由强到弱依次为丁酸、丙酸、乙酸。所有未被机体吸收的 SCFAs 均被排出体外（Ganapathy 等，2010）。如上所述，根据两种转运蛋白在全身的表达情况，提示 SCFAs 的摄取对于机体健康而言具有重要意义。

2 乳短链脂肪酸的检测方法

SCFAs 对机体生理功能和疾病健康都具有重要的作用，因此如何准确测定乳中 SCFAs 引起了众多学者的关注，并在过去 10 年得到了积极的研究。目前，多种检测平台已被应用于乳中 SCFAs 的检测，包括气相色谱（gas chromatography，GC）、液相色谱（liquid chromatography，LC）、近红外光谱（near infrared，NIR）、核磁共振氢谱（nuclear magnetic resonance，NMR）等，乳中 SCFAs 检测方法汇总见表 2.1。但由于乳中 SCFAs 存在形式的特殊性，其检测方法仍然具有局限性。

2.1 气相色谱技术

GC 是一种利用气体作流动相的色谱分离分析方法，具有分离能力强、检测器选择简单、定量准确、灵敏度高以及仪器成本相对低廉等优点。由于 SCFAs 易挥发的特点，GC 被广泛应用于 SCFAs 分析。根据 GC 连接的检测器不同，GC 主要分为气相色谱–氢火焰离子检测器（gas chromatography–flame ionization detector，GC–FID）和气相色谱–质谱联用（gas chromatography–mass spectrometry，GC–MS）两种。GC–FID 主要依据标准品的保留时间对物质进行定性分析，GC–MS 则通过选择目标化合物的特征离子对缺乏标准品的物质进行定性分析，因此 GC–MS 具有更高的检测通量和灵敏度。

Zebari 等（2019）对生乳中脂质进行水提后检测乳中 SCFAs 含量，前处理简单、快速，不足之处在于通过水提法进行 GC 分析时可能因 SCFAs 部分水解而导致回收率降低。此外，水提法并未对生乳中富含的蛋白质等成分进行有效除去，可能会影响 SCFAs 检测的准确性与精密度，也会加快 GC 等仪器和相关配件的损耗（Raterink 等，2014）。Jiang 等（2016）利用 0.5% 盐酸乙醇溶液去除乳中干扰检测的蛋白质，抑制 SCFAs 的水解。硫酸、偏磷酸、甲酸等也可抑制 SCFAs 在水中的电离，达到提高 SCFAs 提取效率和峰形的效果。然而，SCFAs 羧基基团极性较强，在 GC 分析时只进行酸化步骤会出现吸附作用而导致结果重复性较差，特别是低浓度范围（SCFAs < 1 mmol/L）容易产生这种误差（Ghoos 等，1991）。因此，在 GC 分析前可以对 SCFAs 进行衍生化，提高目标分

析物的热稳定性，降低挥发性酸的蒸发损失，提高检测分离度和灵敏度，改善色谱保留时间、峰形等，减少样品与色谱柱涂层间的反应。Dai 等（2020）利用氢氧化钾甲醇皂化后碱催化衍生化生成 SCFAs 甲酯的方法对不同胎龄婴儿摄入的母乳中以甘油三酯形式存在的 C4:0、C6:0 进行检测。Contarini 等（2017）采用相同的碱催化衍生化试剂进行甲酯化的方法对驴乳极性脂质磷脂、神经酰胺和胆固醇等的 SCFAs 组成进行测定。Liu 等（2017）利用硫酸 – 甲醇进行酸催化的甲酯化反应探究生乳 SCFAs 与热应激的关系。David 等（2018）利用庚烷 / 乙醚提取脂质，固相萃取脂质中的游离态 SCFAs，随后利用三氟化硼丁醇进行衍生化反应，生成分子量更高、挥发性更小的 SCFAs 丁酯来测定生牛乳中游离 C4:0、C6:0。丁岩等（2019）通过 α – 溴苯乙酮和 18– 冠 –6 醚衍生化后利用 GC 测定了发酵乳中 7 种 SCFAs。综上所述，由于乳中 SCFAs 存在形式的特殊性以及基质的复杂性，利用 GC 分析乳中 SCFAs 含量的研究是相对有限的，且不同方法中检测到的 SCFAs 种类是较为模糊的。

2.2 液相色谱技术

液相色谱法是以液体为流动相，通过高压输液系统实现两相分离的监测技术，具有高效、灵敏、分离效率高、选择性好、多组分同时定量、适用范围广等优点。由于 SCFAs 易溶于水的性质，液相色谱法也被应用于各种生物样品中 SCFAs 的检测。SCFAs 化学结构中缺乏具有紫外吸收或产生荧光的基团，因此液相色谱检测 SCFAs 需要首先对 SCFAs 进行衍生化。常见的紫外衍生化试剂有苄基、苯甲酰甲基、对硝基苄基、对 – 溴苯甲酰甲基等，常见的荧光衍生化试剂有香豆素、喹啉、重氮甲烷等。Li 等（2022）利用氨水解离乳中乳脂肪球膜释放 SCFAs，并利用乙醇 – 二乙醚 – 石油醚提取脂质，利用 3– 硝基苯肼衍生化，最后利用液相色谱 – 质谱检测生牛乳中 8 种 SCFAs。通过液相色谱分析乳中 SCFAs 组成的研究较少，液相色谱被广泛应用于其他生物基质 SCFAs 的分析。

2.3 波谱技术

波谱技术是利用射频或微波电磁场与物质的共振相互作用来确定物质结构的高新技术，具有无损、快速、重现性高、覆盖范围广、制备工艺简单等优点，被广泛应用于代谢组学分析。波谱技术可直接进行 SCFAs 的分析而无需衍生化操作，但对于乳中微量 SCFAs 的含量检测的准确性较差。Ferrand 等（2014）通过中红外光谱技术和偏最小二乘法回归、遗传算法和偏最小二乘法回归、一阶导数和偏最小二乘法回归结合，获得牛乳和羊乳 SCFAs 含量的良好结果。Wiking 等（2017）通过核磁共振、质子转移质谱法测定乳中与风味和气味相关的 SCFAs 浓度。Prentice 等（2019）利用核磁共振氢谱研究了乳中 SCFAs 与婴幼儿生长发育的关系。Stison 等（2020）利用核磁共振氢谱对 5 个国家（澳大利亚、日本、南非、挪威、美国）的 109 位母亲产后 1 个月母乳样品中 C2:0、C4:0 进行了测定。此外，碳 -13 核磁共振也被作为其他动物特色乳中乳脂肪掺假鉴定的有效手段（Gianluca 等，2013）。

表 2.1　生乳短链脂肪酸分析方法

样品	样品前处理		检测方法		定量方法	检测结果			参考文献
	方法	前处理试剂	仪器	色谱柱		类型	数量	定量限（μg/mL）	
牛乳	水提法		GC–FID		内标法	游离态	7 种	0.50 ~ 1.25	Zebari 等，2019
母乳	酸化法	5% 盐酸乙醇	GC–MS	DB–FFAP	内标法	游离态	5 种		Jiang 等，2016
母乳	甲酯化	氢氧化钾 – 甲醇振荡 2 min	GC–MS	TRACE TR–FAME	外标法	结合态	2 种	0.08 ~ 0.16	Dai 等，2020
牛乳	丁酯化	三氟化硼 – 丁醇 80℃水浴 1 h	GC–MS	CP FFAP CB	外标法	游离态	2 种	20	Mannion 等，2018
发酵乳	衍生法	α– 溴苯乙酮、18–冠醚 –6 100℃水浴 40 min	GC–ECD	DB–23	外标法	游离态	7 种	20	丁岩 等，2019
牛乳	衍生法	3– 硝基苯肼 30℃水浴 30 min	液相色谱 – 质谱	BEH C18 AX	外标法	游离态	8 种	0.01 ~ 0.04	Li 等，2022
牛羊乳			中红外光谱				2 种		Ferrand 等，2014
牛乳			核磁共振			游离态	1 种		Wiking 等，2017
母乳			核磁共振			结合态	3 种		Prentice 等，2019
母乳			核磁共振				3 种		Stinson 等，2020

注：表内空白代表无此项；GC–FID 表示气相色谱 – 氢火焰离子检测器；GC–MS 表示气相色谱 – 质谱；GC–ECD 表示气相色谱 – 电子俘获检测器。

3 短链脂肪酸的生物功能

3.1 短链脂肪酸对肠道健康的影响

肠道每天会产生 500 ～ 600 mmol 的 SCFAs，其中乙酸盐、丙酸盐和丁酸盐是人体中含量最多的 SCFAs，大约以 60∶20∶20 的摩尔比存在于结肠中（Cummings 等，1987），而甲酸盐、戊酸盐和己酸盐的产量较少。SCFAs 在结肠产生后被结肠细胞迅速吸收，主要是通过单羧酸转运蛋白 1（monocarboxylate transporter，MCT1）和钠偶联单羧酸转运蛋白 1（sodium-dependent monocarboxylate transporter，SMCT1）介导的主动转运。MCT1 以 H^+ 依赖性、电中性方式转运 SCFAs，而 SMCT1 通过电生成转运 SCFAs。还有一小部分未解离的 SCFAs 通过被动扩散被结肠细胞吸收，通过进入线粒体的三羧酸循环为细胞产生 ATP 和能量（Schonfeld 等，2016），而未在结肠细胞中代谢的 SCFAs 被输送到门静脉循环。

前期体内和体外试验均证实 SCFAs 在改善肠道健康方面发挥了重要作用（表 3.1）。肠屏障是由上皮细胞通过细胞间的连接构成，促进营养物质的吸收，阻止有害物质和病原体突破肠黏膜和肠上皮屏障。研究表明，SCFAs 能够改善肠道屏障功能，例如丁酸能够刺激小鼠结肠中的黏蛋白生成（Gaudier 等，2009）；增加紧密连接蛋白的表达水平；激活腺苷酸活化蛋白激酶（AMP）/AMP 活化蛋白激酶（AMPK）、诱导调节性 T 细胞（regulatory T cell，Treg）控制炎症来促进上皮屏障功能（董霞等，2022；Silva 等，2020）。而 SCFAs 的缺乏则会导致肠道通透性增加，从而引发炎症级联反应，诱发炎症性疾病的发生（林杨凡等，2022）。因此，SCFAs 主要通过下调炎症细胞因子分泌，调节肠上皮细胞的紧密连接和完整性来促进肠道健康。此外，SCFAs 还可通过肠神经系统影响中枢神经系统（CNS）的炎症，并与迷走神经传入相互作用调节机体健康（表 3.1）。

表 3.1 SCFAs 对肠道健康的影响

模型	SCFAs 剂量和处理时间	影响	参考文献
断奶前或断奶后的仔猪	丁酸钠 3 g/kg，断奶前 24 d/ 断奶后 12 d	↓胃排空和肠黏膜重量	Le 等，2009

续表

模型	SCFAs 剂量和处理时间	影响	参考文献
断奶仔猪	丁酸钠 500 mg/kg、1000 mg/kg，21 d	改善肠道形态，↓近端结肠梭菌和大肠杆菌的总活菌数，↓血清 TNF-α 和 IL-6 水平，↓肠道中 NF-κB 水平	Wen 等，2012
断奶仔猪	丁酸钠 1000 mg/kg，21 d	↓腹泻发生率，↑血清 IgG，↑空肠 IgA^+ 细胞计数	Fang 等，2014
刚断奶的仔猪	三丁酸甘油酯 5 g/kg，4 周	改善肠道形态，↑双糖酶活性	Hou 等，2006
结肠炎猪模型	三丁酸甘油酯 1%，21 d	↓细胞凋亡导致的肠损伤，↑紧密连接形成，↑ EGF 信号传导	Hou 等，2014
C57BL/6 小鼠 T84 细胞	三丁酸甘油酯 5 mmol/L，7 d 丁酸 1 mmol/L、5 mmol/L、10 mmol/L，0 h/6 h/12 h/24 h/48 h	↑上皮屏障功能和伤口愈合	Wang 等，2020
IPEC-J2 细胞	丁酸钠 4 mmol/L，48 h	↑肠道创口的愈合	Ma 等，2012
IPEC-J2 细胞	丁酸 2 mmol/L、4 mmol/L、8 mmol/L、16 mmol/L，24 h	↑宿主防御肽基因表达	Zeng 等，2013
全肠外营养新生仔猪	SCFA（乙酸、丙酸、丁酸：36 mmol/L、15 mmol/L、9 mmol/L）；急性（4 h/12 h /24 h）和慢性（3/7 d）	↓肠萎缩，↑肠细胞增殖，↓肠细胞凋亡，↑肠道适应能力	Bartholome 等，2004
断奶仔猪	SCFA（乙酸、丙酸、丁酸：20.04 mmol/L、7.71 mmol/L、4.89 mmol/L；40.08 mmol/L、15.41 mmol/L、9.78 mmol/L），7 d	改善肠道形态、↓凋亡细胞百分比和维持肠道屏障功能	Diao 等，2019
无菌仔猪	SCFA（乙酸、丙酸、丁酸：45 mmol/L、15 mmol/L、11 mmol/L），21 d	↑空肠和血清中 GLP-2 的含量，↓回肠、结肠中 IL-1β、IL-6 水平，↑肠道发育和吸收功能，↑肠道免疫功能	Zhou 等，2020
斑马鱼仔鱼炎症模型	SCFA（乙酸、丙酸、丁酸：3 mmol/L、6 mmol/L、9 mmol/L），7 d	↑仔鱼的存活率，缓解炎症，↓促炎细胞因子的表达水平	Morales 等，2021

注：↑表示升高，↓表示降低，下同。

SCFAs能够通过多途径维持和改善肠道健康。例如，SCFAs可以通过促进婴儿肠道菌群的定植，促进肠组织的生长和成熟（Liu等，2016）；保持肠道屏障的完整性，防止肠道炎症（Lewis等，2010）。SCFAs主要通过激活腺苷酸单磷酸（adenosine monophosphate，AMP），活化腺苷酸激活蛋白激酶（the AMP-activated protein kinase，AMPK）上调紧密连接蛋白的表达来增强肠屏障功能（D'souza等，2017；Wells等，2017）。SCFAs还可通过改变肠上皮细胞系O_2的消耗，稳定屏障保护相关的缺氧诱导因子（Hypoxia inducible factor，HIF），从而增强肠道屏障功能（Kelly等，2015）。

研究表明，富含乙酸盐的酸奶可以改善肠道上皮对肠道的保护功能（Chang等，2021）。将巴马母猪生产的12头新生无菌仔猪随机分为对照组和SCFAs处理组，对照组每天口服25 mL/kg无菌生理盐水，SCFAs组每天口服25 mL/kg无菌SCFAs混合物（乙酸、丙酸和丁酸分别为45 mmol/L、15 mmol/L和11 mmol/L），结果表明SCFAs显著降低了空肠、回肠和结肠中促炎细胞因子白细胞介素（Interleukin-1β，IL-1β）和IL-6的mRNA丰度，显著提高了空肠和血清中胰高血糖素样肽（glucagon-like peptide 2，GLP 2）的含量，并增加了血液中白细胞、中性粒细胞和淋巴细胞的计数，且揭示外源性补充SCFAs（独立于肠道菌群产生的SCFAs）可通过促进肠道发育和吸收功能以及增强肠道免疫功能，从而改善肠道健康（Zhou等，2020）。此外，SCFAs还影响胃肠道的黏液生成，从而减少上皮细胞与管腔微生物和有毒物质之间的相互作用，保护细胞免受消化过程中pH值波动的影响（Pelaseyed等，2014）。乙酸盐和丁酸盐主要通过上调黏蛋白2（mucoprotein2，MUC2）基因表达刺激和增加大鼠结肠中黏蛋白的分泌和生成（Barcel等，2000；Gaudie等，2009）。

除此之外，SCFAs还能缓解肠道疾病。在肠易激综合征（irritable bowel syndrome，IBS）模型组中，从出生后第10天开始，对模型小鼠注射SCFAs（第10～15天，0.3 mL/d；第16～21天，0.5 mL/d），在最后一次注射SCFAs 2周后，进行内脏超敏反应和结肠运动试验，结果表明0.5 mmol/L、1 mmol/L、5 mmol/L、10 mmol/L、30 mmol/L SCFAs均能诱导IBS与腹泻相关的结肠张力、自发收缩幅度和频率以剂量依赖性降低，其中乙酸钠、丙酸钠、丁酸钠发挥抑制作用的半最大效应浓度（concentration for 50% of maximal effect，EC50）分别为 4.79 mmol/L、7.58 mmol/L 和 3.15 mmol/L，丁酸抑制效果最好（Shaidullov 等，2021）。一项旨在分析 3 种联合 SCFAs 处理对 2,4,6– 三硝基苯磺酸（2,4,6-trinitrobenzenesulfonic acid，TNBS）诱导的斑马鱼仔鱼肠道炎症影响的研究显示，SCFAs 显著提高了 TNBS 处理后幼仔的存活率，保护了肠道内吞功能，减少了炎症细胞因子的表达并减少了肠道中由 TNBS 引起的中性粒细胞浸润（Morales 等，2021）。此外，丙酸直接作用于 T 细胞亚群（γδT）细胞，通过抑制 HDAC 活性来抑制炎症性肠病患者 γδT 细胞产生 IL–17（Dupraz 等，2021）。同样，另一项研究也表明丁酸能通过 HDAC8 抑制己糖激酶 2 的表达，起到缓解结肠炎的作用（Hinrichsen 等，2021）。此外，丁酸还可通过 HDAC8 在上皮细胞系和小鼠结肠内诱导肌动蛋白结合蛋白突触素（recombinant synaptopodin，SYNPO）的表达，促进肠道上皮屏障功能和伤口愈合（Wang 等，2020）。黏蛋白缺乏会加剧各种肠道疾病，例如黏膜炎，而 SCFAs 能促进黏蛋白的合成来减少肠道损伤，减少溃疡；与 SCFAs 混合溶液相比，单独使用丁酸能更有效地改善肠道损伤参数，缓解肠道黏膜炎（Ferreira 等，2012），因此丁酸可作为结肠黏膜的抗炎剂。

总而言之，SCFAs 可以通过激活或上调 AMPK、增加肠道屏障相关蛋白的表达来提高紧密连接蛋白和黏蛋白表达水平，促进肠道屏障的完整性，进而维持和促进肠道的健康。此外，SCFAs 还可通过抑制 HDAC 活性来缓解肠道相关疾病。

3.2 短链脂肪酸对脑发育的影响

3.2.1 短链脂肪酸可改善大脑的认知功能

前期研究已经表明 SCFAs 不仅可以改善认知功能，还可以改善脑源性疾病。作为肠道菌群的代谢物，SCFAs 缺乏的小鼠模型表现出认知障碍，而补充 SCFAs 能够改善小鼠的认知功能（Shi 等，2021）。母亲饮食中的纤维能够通过改变 SCFAs 水平来调节后代的神经认知功能（Yu 等，2020）。此外，丁酸可降低脂多糖诱导的大鼠原代小胶质细胞、海马组织培养物以及小脑颗粒神经元、星形胶质细胞和小胶质细胞共培养物的炎症反应（Huuskonen 等，2004）。通过单剂量三乙酸甘油酯（6 g/kg）补充乙酸盐能减少大鼠神经炎症的发生（Soliman 等，2011）。在帕金森病动物模型中，丁酸处理改善了试验动物的运动障碍和多巴胺缺乏症状（Paiva 等，2017；Sharma 等，2015）。在躁狂症的动物模型中，丁酸钠逆转了大鼠的多动行为，降低了前额叶皮质、海马和杏仁核中线粒体呼吸链复合体的活性，并逆转了抑郁和躁狂行为（Resende 等，2013）。与单独使用氟西汀（一种抗抑郁药）相比，丁酸盐（0.6 g/kg）联合氟西汀（10 mg/kg）显著减少了小鼠的抑郁行为，且提高了脑源性神经营养因子（brain-derived neurotrophic factor，BDNF）的表达水平（Schroeder 等，2007），推测 BDNF 可能在丁酸抗抑郁行为中发挥了重要的作用。在阿尔茨海默病小鼠模型中，丁酸处理恢复了小鼠的记忆功能，并提高了其与学习相关基因的表达水平（Govindarajan 等，2011）。此外，在小鼠体内全身注射（1.2 g/kg）和海马内注射（55 mmol/L）丁酸钠可持续消除小鼠的恐惧（Stafford 等，2012）。在自闭症谱系障碍小鼠模型中，腹腔注射 100 mg/（kg・d）丁酸钠连续 10 d 可通过调节兴奋 / 抑制平衡（E/I 平衡）来增加小鼠的社会行为（Kratsman 等，2016）。前期研究证实 SCFAs 水平与婴幼儿的神经认知功能息息相关（Yu 等，2020），但目前关于 SCFAs 缓解脑源性疾病的作用大多集中在动物模型上，还缺乏相关的人类研究数据来支持这一结论。

由于 SCFAs 转运蛋白 MCT 在内皮细胞上的大量表达，使得 SCFAs 可以穿过血脑屏障（blood brain barrier，BBB）到达大脑（Mitchell 等，2011）。颈动脉注射 14C-SCFA 后，大鼠脑内通过 BBB 摄取 SCFAs 的相对顺序为丁酸、丙酸和乙酸（Oldendorf 等，1973）。在人脑组织中，丙酸盐和丁酸盐的平均浓度分别为 18.8 pmol/mg 和 17.0 pmol/mg，用正电子发射计算机断层扫描成像显示，给大鼠静脉注射 11C- 乙酸盐后，约 3% 立即被大脑吸收，结肠注射 20 min 后，约 2% 被大脑吸收（Frost 等，2014）。

尽管 SCFAs 对 CNS 作用的确切机制仍不清楚，但大量体内、体外研究均表明（表 3.2），其对关键的神经和行为过程产生广泛影响，并可能参与神经发育。同时，SCFAs 通过调节神经免疫，改善神经元功能，具有缓解或治疗神经退行性疾病和脑源性疾病的潜力（Dinan 等，2017；Kelly 等，2017）。SCFAs 与营养物质从血液循环到大脑的通道密切相关，可激活小胶质细胞，促进神经元生长和突触可塑性，在脑发育方面发挥着重要作用。SCFAs 能促进神经细胞的有丝分裂，提高生长速度，调节早期神经系统的发育（Yang 等，2020），这也为研究 SCFAs 如何调节动物早期神经系统发育提供了线索。SCFAs 已被证实对神经退行性疾病具有重要的保护作用，通过移植富含 SCFAs 的盲肠粪便细菌并同时补充丁酸能有效缓解缺血性脑中风（Chen 等，2019）。体内、体外研究均表明，丁酸盐可诱导小胶质细胞的形态和功能改变，使其趋于稳态，并抑制脂多糖诱导的促炎因子的表达以及抑郁和躁狂症状（Yamawaki 等，2018；Wang 等，2018）。同样，乙酸盐也能够调节星形胶质细胞的炎性因子的表达水平和相关信号通路，从而减轻大鼠神经炎症模型中神经胶质细胞的激活（Soliman 等，2012）。丁酸盐能抑制大鼠小胶质细胞、海马组织以及星形胶质细胞的炎症反应（Huuskonen 等，2004）。此外，丁酸可以促进神经元可塑性、增强记忆、恢复认知损伤，减少神经毒性、神经炎症和行为异常，且通过抑制组蛋白去乙酰化酶（histone deacetylase，HDAC）缓解多种 CNS 疾病症状（Patnala 等，2017；Sharma 等，2015）。因此，SCFAs 可能通过调节神经发育和小胶质细胞的激活来影响试验动物的大脑功能。

表 3.2　SCFAs 对大脑功能的影响

模型	SCFAs 剂量和处理时间	作用	参考文献
星形胶质细胞	乙酸钠 12 mmol/L，1 h/2 h/4 h	↓ TNF-α 和 IL-1β 表达水平，↓ p38 MAPK、JNK 和 NF-κB 磷酸化	Soliman 等，2012
SD 大鼠	三乙酸甘油酯 6 g/kg，0.5 h/1 h/2 h/4 h/24 h	抑制 HDAC2，↑脑组蛋白的乙酰化，↓神经炎症	Soliman 等，2011
C57BL/6J 小鼠	丁酸钠 0.6 g/kg，28 d	↑海马和额叶皮层的组蛋白高度乙酰化，↑学习和记忆力，发挥抗抑郁作用	Schroeder 等，2007

续表

模型	SCFA 剂量和处理时间	影响	参考文献
脑萎缩小鼠模型野生型小鼠	丁酸钠 1.2 g/kg，4 周	抑制 HDAC，↑野生型小鼠和脑萎缩小鼠的学习和记忆力	Fischer 等，2007
C57BL/6 小鼠	丁酸钠 1.2 g/kg，7 d	改变海马的基因表达，↑抗抑郁作用	Yamawaki 等，2018
SD 大鼠	丁酸 30 mg/kg，14 d	↓脑缺血大鼠的神经功能损伤，↓脑梗死体积，↓脑水肿	Chen 等，2019
人脑血管内皮细胞系	丙酸 1 μmol/L，24 h	↓ BBB 受氧化应激的程度	Hoyle 等，2018
C57BL/6 小鼠	SCFAs(丙酸钠、丁酸钠、乙酸钠：25 mmol/L、40 mmol/L、67.5 mmol/L)，4 周	↑小胶质细胞的稳态、发育	Erny 等，2015
C57BL/6J 小鼠	SCFAs（乙酸盐、丙酸盐、丁酸盐：67.5 mmol/L、25 mmol/L、40 mmol/L），7 d	↓社会应激带来的压力，抵消社会应激的持久性影响，↑抗抑郁、抗焦虑作用	Van de Wouw 等，2018
老龄缺血性卒中小鼠模型	含有 SCFAs 的粪菌移植	↓老年中风小鼠的神经功能缺损和炎症	Lee 等，2020

3.2.2 短链脂肪酸在肠 – 脑轴方面的作用

饮食中膳食纤维的缺乏破坏了肠道屏障，减少了 SCFAs 的产生，进一步研究 SCFAs 受体敲除后对肠道和大脑功能的影响，发现 SCFAs 受体失活介导肠功能障碍和认知障碍，这进一步证实 SCFAs 在肠道功能和认知障碍之间起着关键作用（Shi 等，2021）。对 SCFAs 浓度和脑组织抗氧化能力以及肠道炎症因子之间进行相关性分析发现，乙酸盐、丙酸盐和丁酸盐均与肠道促炎细胞因子 TNF-α 呈显著负相关，但与抑炎细胞因子 IL-10 呈显著正相关，乙酸盐和丁酸盐与脑组织总抗氧化能力呈显著正相关（Ren 等，2022）。生理浓度的 SCFAs 通过母体肠–子代脑轴促进动物幼体的神经发育（Yang 等，2020）。中风后会改变肠道菌群的定植，Sadler 等（2020）表明 SCFAs 是肠道–免疫–脑轴的介质，SCFAs 通过循环淋巴细胞激活小胶质细胞，是中风后神经元可塑性的有效促再生调节剂，也证实 SCFAs 是中风后肠脑轴上缺失的关键一环。脑缺血会导致肠道菌群失调、肠道通透性增加以及肠道屏障破坏，丁酸可以通过修复渗漏的肠屏障显著降低肠漏，显著改善脑缺血、脑卒中（Chen 等，2019）。SARS-CoV-2 感染后与受体结合导致重要营养成分的转运减少、肠道功能失调、肠道血液屏障通透性增加以及全身炎症水平升高，而高纤维饮食或补充 SCFAs 可以通过肠脑轴预防或减轻 COVID-19 引起的肠道和神经损伤（Sajdel-sulkowska 等，2021）。

FFAR2 和 FFAR3 是 SCFAs 的特异性受体，也是包括肠道、大脑在内的许多器官中发现的 SCFAs 的特异性受体（Kimura 等，2020）。CNS 和周围神经系统（peripheral nervous system，PNS）中均存在功能性 SCFAs 受体 FFAR2 和 FFAR3（Dalile 等，2019）。SCFAs 通过与 GPRs 的相互作用影响肠脑通讯功能，例如 SCFAs 结合 FFAR3 后，通过迷走神经抑制食欲（Bonaz 等，2018）。丁酸通过环腺苷酸磷酸二酯酶（cAMP）依赖性直接激活肠道糖异生基因表达，而丙酸盐作为肠道糖异生的底物，通过 FFAR3 介导的肠 - 脑神经回路间接激活肠道糖异生基因表达（De vadder 等，2014）。这些研究表明，SCFAs 通过结合 GPRs 作用于门静脉周围传入神经系统，随后向 PNS 和 CNS 区域发出信号，以诱导肠道功能的改变。此外，SCFAs 还可通过抑制 HDAC 发挥神经保护作用，HDAC 的抑制会触发基因表达发生改变，影响免疫细胞的分化和功能以及上皮细胞的表观遗传调控，进而调节肠道和大脑功能（Obata 等，2018）。

3.3 短链脂肪酸对肾脏健康的影响

肠道微生物群产生的 SCFAs 被吸收到血液中，并由 GPR41、GPR43、GPR109A 和 Olfr78 等 G 蛋白耦合受体（GPRs）吸收后到达肝脏、脂肪组织和肾脏等较远的组织。不同程度地激活 GPRs；丙酸盐激活 GPR41 和 GPR43，乙酸盐激活 GPR43，丁酸盐激活 GPR41（图 3.1）。

Olfr-78、GPR-43、GPR-41 和 GPR-109A 均在肾脏中表达（Pluznick 等，2013；Flegel 等，2013；Zhang 等，2013），因此 SCFAs 可以影响肾

脏健康。Olfr-78 在肾球旁器和小阻力血管的平滑肌细胞中表达（Pluznick 等，2013）；其人类直系同源物（hOR51E2）在人类肾脏以及心脏和骨骼肌等多种其他组织中表达（Flegel 等，2013）；GPR-43 在人胚胎肾细胞中表达；GPR-41 也在肾动脉和平滑肌细胞中表达，这表明这些受体的生物学作用可能超出了它们在调节肠内分泌细胞释放肽激素中的作用，包括调节肾脏健康（Bolognini 等，2015）。GPR109A 在足细胞中的表达增加，通过稳定肾小球基底膜足细胞和减轻肾小球硬化和肾脏炎症，在肾损伤的动物模型中具有减少蛋白尿发生的作用（Felizardo 等，2019）。因此，肠道微生物群产生的 SCFAs 对肾脏健康和疾病防护具有潜在作用。

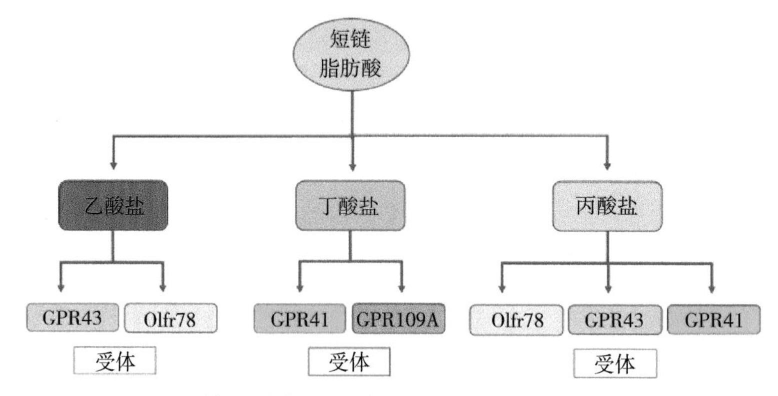

图 3.1 SCFAs 受体及其与不同类型 SCFAs 的关系（Zaky 等，2021）

补充丁酸可能通过上调急性肾损伤（AKI）的抗氧化和抗炎作用，减轻大鼠肾脏缺血 / 再灌注（I/R）损伤。此外，临床数据显示丁酸水平升高与肾移植后肾功能恢复相关（Sun 等，2022）。同样，醋酸盐和醋酸盐产生菌也减少了肾脏 I/R 损伤（Andrade-Oliveira 等，2015）。通过降低参与 NF-κB 信号通路的两个因素 IL-6 水平和氧化应激，丁酸盐被证明可以改善肾功能（Machado 等，2012）。此外，丁酸盐通过其抗氧化能力降低了肾脏毒性（Sun 等，2013）。与健康对照组相比，慢性肾脏疾病（CKD）患者的丁酸水平较低，这表明补充丁酸盐可以减缓 CKD 的进展（Wang 等，2019）。此外，丙酸可以通过 FFAR2 和 FFAR3 信号通路部分抑制肾功能衰竭（Mikami 等，2020）。值得注意的是，研究表明，高纤维饮食可以产生更多的短链脂肪酸，对慢性肾脏病有益（Kieffer 等，2016；

Vaziri 等，2014）。单链脂肪酸也被认为是延缓慢性肾脏病进展的潜在药理学候选物（Felizardo 等，2019）。

除了 AKI 和 CKD 外，糖尿病肾病（DKD）的研究也逐渐活跃起来。据报道，DKD 患者血清和粪便中的 SCFAs 水平较低（Zhong 等，2021）。通过 FFAR2 和 GPR109A 信号通路，高纤维饮食和单链脂肪酸（醋酸盐、丁酸盐和丙酸）保护糖尿病小鼠免受糖尿病肾病的影响。在体外，SCFAs 还减少了促炎细胞因子的基因表达，证明 SCFAs 具有预防糖尿病肾病的能力（Li 等，2020）。SCFAs 给药可减少肾结晶，与对照组相比，肾结石组产生 SCFAs 的细菌数量以及与 SCFAs 产生相关的代谢途径减少（Liu 等，2020）。此外，SCFAs 治疗可能成为治疗肠源性炎症反应所致肾损伤的新靶点（Huang 等，2017）。

然而，研究发现，超临界脂肪酸对肾功能有一定的有害影响。乙酸盐（C2）诱导的肾脏疾病，也称为 C2RD 病，是一种 T 细胞介导的肾脏疾病，伴有进行性输尿管炎和肾积水。研究表明，这是由于长期服用高于生理水平的短链脂肪酸（200 mmol/L 醋酸盐、200 mmol/L 丙酸盐和 200 mmol/L 丁酸盐）导致的（Park 等，2016）。此外，喂食高膳食纤维饮食的小鼠血液中丁酸盐水平更高，大肠杆菌种类更少，这使它们更容易受到大肠杆菌的影响，肾脏中的 Gb3 蛋白水平增加，最终导致严重的肾脏损害（Zumbrun 等，2014，2013）。

综上所述，SCFAs 通过改变肠–肾轴上的受体及相关信号通路来调节肾功能。SCFAs 对肾功能有双重作用，包括降低 AKI、CKD 和 DKD 的严重程度（正调节）以及增加肾积水（C2RD 病）的炎症和细菌易感性（负调节）。尾静脉注射 SCFAs 对肾脏疾病具有保护作用，这与高纤维饮食可能通过肠 – 肾轴对肾脏产生有益作用，可在循环系统中产生更多的 SCFAs 的结果一致。

3.4 短链脂肪酸对肝脏健康的影响

在全球范围内，慢性肝病是最常见的疾病之一，约有 8.4 亿人患有慢性肝病，每年有 200 万人死于慢性肝病（Byass，2014）。慢性肝病的特点是肝功能逐渐恶化，包括凝血因子和其他蛋白质的生成、代谢有害产物的解毒和胆汁的排泄。在慢性肝病中，肝实质的炎症、破坏和再生的持续过程会导致肝纤维化和肝硬化。最近的研究发现，SCFAs 可以预防和控制多种肝脏疾病，特别是非病毒性肝病（Li 等，2021；Meng 等，2018）。非病毒性肝病主要涉及非酒精性脂肪性肝病（NAFLD），酒精性肝病（ALD）以及药物或污染物等引起的肝损伤。

NAFLD 包括单纯性脂肪变性、非酒精性肝炎（NASH）和肝硬化，已成为世界上发病率最高的肝病（Loomba 等，2013）。它已成为欧美等发达国家和中国富裕地区慢性肝病的重要诱因，可严重增加肝细胞癌的发病风险（Behary 等，2021；Wong 等，2014）。丁酸盐可逆转 NASH 的发展，这部分与肠道微生物组的变化有关（Ye 等，2018）。

根据已报道的研究，可总结出 SCFAs 预防 NAFLD 的几种可能作用途径。首先，SCFAs 对肝脏脂肪变性有影响。在小鼠中，喂食 SCFAs 会使脂质氧化增加 3 倍，从而以 PPARγ 依赖的方式减少肝脏脂质合成（Den Besten 等，2015）。小鼠经胃给予丁酸盐后，肝脏中的脂质沉积明显减少，炎症活动减弱（Jin 等，2015a）。通过抑制黄嘌呤氧化酶的活性，补充乙酸盐可保护大鼠免受尼古丁引起的肝脏脂质过剩的影响（Dangana 等，2019）。在兔的肝脏中，醋酸盐可以促进脂肪分解和脂肪酸氧化，进而防止脂质积累（Liu 等，2019）。醋酸盐可抑制乳糜微粒的分泌，导致流入循环系统的脂质减少，进而缓解非酒精性脂肪肝（Araújo 等，2020）。在临床试验中，研究发现，与健康对照组相比，患有非酒精性脂肪肝的成人体内产生 SCFAs 的肠道微生物（类杆菌）数量减少（Mouzaki 等，2013）。因此，SCFAs 和作为 SCFAs 来源的高纤维饮食可缓解肝病（De 等，2012）。

其次，SCFAs 可降低胰岛素抵抗。由于肝脏是负责储存和代谢葡萄糖的器官，肝脏对葡萄糖的低效利用会导致胰岛素抵抗（Zhang 等，2022）。肝脏中的 SCFAs 可通过 AMPK 信号通路调节肝脏的胰岛素敏感性。此外，SCFAs 可与肝细胞中的 GPR41 和 GPR43 受体结合，从而抑制 AMPK 依赖性葡萄糖生成（Den Besten 等，2015）。丙酸盐通过与肝脏 GPR43 结合并激活 AMPK 信号通路抑制葡萄糖生成（Yoshida 等，2019）。

最后，SCFAs 可增强肠道通透性。根据体内研究结果，高脂饮食诱导的肝脂肪变性与较高的肠道通透性有关（Cani 等，2007）。此外，高脂饮食模型也揭示了非酒精性脂肪肝与肠道通透性之间的关系；具体而言，当 TNBS 诱发结肠炎时，脂肪性肝炎的严重程度会增加，这一点可通过肝酶水平和非酒精性脂肪肝活动评分（NAS）证实（Mao 等，2015）。鉴于 SCFAs 可逆转肠道通透性，如上所述，可以推测 SCFAs 可通过保护肠道屏障来预防 NAFLD。

除了直接影响外，SCFAs 对肠道屏障介导的 NAFLD 的潜在影响也值得关注。SCFAs 可以通过改变肠道微生态来保护肠道屏障，从而延缓 NAFLD 相关疾病的发展。在高脂肪饮食诱导的 NAFLD 小鼠模型中，丁酸钠的膳食补充剂增加了肠道菌群中 *Lactobacillus* 的丰度，形成了良性循环和更多的丁酸（Zhou 等，2017）。丁酸可以提高 ZO-1 在小肠中的表达水平，治愈 NAFLD 患者的肠黏膜损伤，从而进一步阻止肠道内毒素向肝脏迁移，最终形成治疗 NAFLD 的良性循环（Zhou 等，2017）。在蛋氨酸胆碱缺乏症饮食（MCD）诱导的非酒精性脂肪性肝炎（NASH）小鼠模型中，补充丁酸盐显著增加了肠道菌群 *Akkermansia* 和 *Roseburia* 等的含量（Ye 等，2018）。一些研究表明 *Akkermansia* 和 *Roseburia* 具有改善肠道屏障和预防免疫系统疾病的功能（Patterson 等，2017）。因此，丁酸盐可以诱导肠道微生物组的保护性转移，修复小鼠肠道屏障，有效预防 MCD 饮食引起的 NASH 相关肝损伤（Ye 等，2018）。补充丁酸钠还可以保护肠道屏障，限制内毒素位移，减缓富含脂肪、果糖和胆固醇的流质饮食诱导的小鼠 NAFLD 的发生，这是通过直接增加十二指肠中紧密连接蛋白的表达来实现的（Cheng 等，2015b）。迄今为止，虽然临床试验证据不足，但不可否认的是，SCFAs 可以通过调节黏液分泌、肠上皮紧密连接、微生物稳态等机制，减少肠道内毒素向肝脏的扩散，从而降低肝脏炎症水平和氧化压力，减缓 NAFLD 的发展。

酒精性肝病（ALD）可由长期大量饮酒引起，乙醇在肝脏中转化为乙醛，乙醛进一步代谢成醋酸盐，此后迅速释放到血液中。此外，肝脏中的乙醇代谢也会产生高活性分子片段，导致氧化应激和肝损伤（Lang 等，2020）。酒精性肝损伤的类型包括脂肪肝（Sehrawat 等，2020）、纤维化和酒精性肝炎，这些病变可单独、同时或连续发生于同一患者（Louvet 等，2015）。SCFAs 的重要作用是为肠上皮细胞提供能量和营养，并保持肠道屏障的完整性和黏膜免疫耐受。长期乙醇给药不仅会减少肠道细菌中 SCFAs 的生物合成，还会减少饱和长链脂肪酸（LCFAs）的生物合成，从而导致肠道屏障功能障碍。据报道，补充 LCFAs 可维持小鼠肠道生化并减少乙醇诱导的肝损伤（Hosomi 等，2022）。一项研究发现，丙酸盐通过维持肠上皮屏障功能和抑制肝 Toll 样受体 4（TLR4）–NF- κB 通路，缓解乙醇诱导的肝脂肪变性并增强肝功能（Xu 等，2022）。此外，一项动物研究显示，戊酸小球球菌通过逆转肠道微生物群失调、调节 SCFAs 代谢来缓解乙醇诱导的肝损伤（Jiang 等，2020）。

至于药物或污染物引起的肝损伤，SCFAs 也发挥了保肝作用。肝脏中的细胞色素 p450（CYP）成熟对代谢活性和异生素解毒很重要。一项体外研究表明，乙酸盐、丙酸盐和丁酸盐的混合物增加了 CYP3A4 和 ALB 在人诱导的多能干细胞来源的肝脏类器官中的表达，从而改善了肝脏功能，增强了肝脏代谢活性和异生解毒功能（Mun 等，2021）。此外，据报道，乙酸盐降低了天冬氨酸氨基转移酶和碱性磷酸酶的血清水平，这表明它改善了肝功能。同时，乙酸盐提高了线粒体效率和三磷酸腺苷的产生（Sahuri–Arisoylu 等，2016）。

然而，SCFAs 对肝脏的影响也具有两面性。有报道称，微生物群产生的过量醋酸盐可能会促进肝脏的脂肪生成。肠道微生物群产生的过量 SCFAs 已被证明是一种额外的能量来源，可能会导致肝脏脂肪的积累（Murugesan 等，2017）。除乙酸盐外，一项动物调查也显示，丙酸盐合成较高会促进肝脏脂肪生成（Gao 等，2019）（表 3.3）。

简而言之，SCFAs 在非病毒性肝病中的作用机制包括调节肝脏脂肪变性、胰岛素抵抗和肠道通透性，从而调节肝功能。在模拟肝病的饮食中添加 SCFAs 可能具有保护作用，这一发现表明，定期在膳食中添加 SCFAs 具有重要意义。

表 3.3　SCFAs 对肝脏健康的作用研究

SCFAs 类型	模型 / 受试者	剂量和持续时间	结果	影响类型	参考文献
丁酸	C57BL/6J 小鼠非酒精性脂肪肝（NAFLD）模型	丁酸钠（0.12 g/mL），6 周	↓ NASH 在保护性肠道微生物组和代谢组调控下的发病率	有益效果	Ye 等，2018
短链脂肪酸	雄性 C57Bl/6J 小鼠，PPARg lox/lox 小鼠	在日粮中添加 5%（重量比）的乙酸盐、丙酸盐或丁酸盐	↓ 肝脏中 PPAR γ 的表达；↓ 肝脏甘油三酯浓度	有益效果	Den Besten 等，2015
丁酸	C57BL/6J 小鼠非酒精性脂肪肝（NAFLD）模型	丁酸钠 [0.6 g/（kg 体重 · d）]，6 周	↓ 肝脏中的炎症；↓ NASH	有益效果	Jin 等，2015
乙酸	Hyla 兔	乙酸盐 [2 g/（kg 体重 · d），每天注射一次]，4 d	↓ 肌内甘油三酯水平；↑ 脂肪酸吸收；↑ 脂肪酸氧化	有益效果	Liu 和 Fu，2019
丙酸	人类 HepG2 肝细胞	丙酸盐（0 mmol/L、0.25 mmol/L、0.5 mmol/L），用于 HepG2 细胞	与肝脏 GPR43 结合并激活 AMPK 信号通路，↓ 葡萄糖生成	有益效果	Yoshida 等，2019
丁酸	初产纯种雌性 SD 大鼠	1 % 丁酸钠饮食	↑ 母体脂肪分解，↑ 脂肪酸吸收和脂质在后代肝脏中积累	负面效果	Zhou 等，2016

3.5　短链脂肪酸对肺健康的影响

肺和消化道之间存在双向连接，称为肠肺轴（Budden 等，2017）。该轴与肺部疾病有关，已知肺部疾病会影响消化系统，反之亦然（Keely 等，2014；Fricker 等，2018）。这种器官间连接是由肠道微生物组实现的，肠道微生物组通过发酵抗性淀粉、果胶和纤维素等膳食纤维（Parada Venegas 等，2019），产生多种代谢物，特别是 SCFAs。SCFAs 是嵌入免疫调节功能的抗炎化学物质，包括抑制趋化性和免疫细胞黏附，以及诱导抗炎细胞因子的表达和刺激免疫细胞凋亡（Ratajczak 等，2019）。肠道问题患者的 SCFAs 浓度降低，使其更容易患肺部疾病（Dang 等，2019）。

据报道，SCFAs 在肺部具有抗炎作用，通过减少肺巨噬细胞和单核细胞产生的 CXCL1 表达，防止气道过度浸润（Aurélien Trompette 等，2018）。除了肺部免疫防御作用外，SCFAs 还通过增加 TJ 蛋白的表达来维持气道上皮屏障，从而保护肺部（Richards 等，2020）。作为最常见的呼吸系统疾病之一，慢性阻塞性肺疾病（COPD）的预防和治疗引起了众多研究人员的关注。SCFAs 对慢性阻塞性肺病的保护作用不容忽视。肺气肿是慢性阻塞性肺病的典型症状，实验动物模型显示营养缺乏和肺泡组织破坏之间存在联系，可导致肺气肿。高纤维饮食可减轻吸烟暴露肺气肿小鼠相关的病理变化，部分原因是 SCFAs（包括乙酸盐、丙酸盐和丁酸盐）的产生增加（Jang 等，2021）。此外，在大鼠缺氧模型和大鼠微血管内皮细胞中，丁酸盐处理可减少肺泡中 $CD68^{+}$ 的聚集以及肺间质中 CD68+ 和 $CD163^{+}$ 肺巨噬细胞的聚集（Karoor 等，2021）。肺气肿与肺泡间隔血管化有关，这可能是由肺泡上皮细胞和内皮细胞凋亡引起的（Petrache 等，2013）。因此，SCFAs 可通过保护内皮细胞和维持肺免疫平衡来预防肺气肿。

支气管微生物群的结构紊乱与慢性阻塞性肺病的恶化有关，微生物在支气管中的定植与肺部免疫功能紧密相关。现有数据表明，肠道和肺部微生物群之间存在一些联系，饮食可改变肠道微生物区系和呼吸道微生物区系（Madan 等，2012）。SCFAs 已被证实可直接影响微生物并改变其毒力（Machado 等，2021）。

SCFAs 也被认为是肠道微生物群和哮喘病因之间的重要联系，其目前已成为减少哮喘患者肺部促炎反应的一种候选疗法（Ríos-Covián 等，2016）。肠道微生物群产生的 SCFAs 可通过诱导 DNA 甲基化和组蛋白修饰的变化，起到表观遗传调节剂的作用（Lee，2019）。SCFAs 可调节宿主的免疫平衡，而宿主的免疫平衡是结肠调节性 T 细胞（Tregs）生长所必需的。调节性 T 细胞功能失调会导致免疫系统调节过度活跃，从而使 Th2 的反应能力崩溃，进而发展为过敏性哮喘（Zhao 等，2018）。丁酸盐通过与 GPCRs（如 FFA 受体 FFA2 和 FFA3）结合，增强 Treg 和 DC 的分化，从而发挥其作为 HDAC 抑制剂的活性（Kespohl 等，2017）。此外，丁酸盐还能通过限制肠道中 NF-κB 介导的 B 细胞活化来缓解肠道炎症（Indrio 等，2017）。研究发现，丁酸盐可通过阻断 GATA 结合蛋白 3（GATA3）和其他促炎细胞因子在人肺第 2 组先天性淋巴细胞（ILCs）中的产生而发挥抗哮喘活性（Lewis 等，2019）。母体肠道细菌产生的短链脂肪酸会间接影响 T 细胞，从而为后代创造一种耐受性免疫环境。GPR41/FFA3、GPR43/FFA2 和 GPR109a 是 SCFAs 与之结合的 3 种 G 蛋白偶联受体。通过与 GPR43 相互作用并直接抑制 HDAC，SCFAs 对中性粒细胞和嗜酸性粒细胞有凋亡作用。研究证明，丁酸盐和丙酸盐可通过降低 T 淋巴细胞的活化来单独调节树突状细胞的活性（Millard 等，2002）。此外，丁酸盐还能激活烟酸受体 GPR109a，缺乏 GPR109a 的小鼠产生 IL-10 的 T 细胞较少（Singh 等，2014）。调节性 T 细胞可产生保护性 Th1 表型和耐受性免疫特征，有助于预防哮喘的发生。在过敏性气道炎症的小鼠模型中，引入的 Treg 可减少现有疾病并防止疾病进展（Thorburn 等，2010）。Treg 的数量和功能受到 SCFAs 的直接影响。乙酸酯和丙酸酯处理后，这些细胞中 Foxp3 和 IL-10 的表达增加，结肠 Treg 的形成也随之增加（Smith 等，2013）。SCFAs 直接和间接影响 T 细胞，产生耐受性免疫特征，Treg 数量和活性增加（Rooks 等，2016）。Trompette 等（2014）和 Thorburn 等（2010）研究表明，母体膳食中碳水化合物的增加与后代过敏性气道炎症严重程度的降低有关。这些研究表明，母亲膳食中的高水平碳水化合物，以及暴露于高浓度的 SCFAs 与后代患哮喘疾病之间存在负相关关系，这些发现为减轻后代哮喘等呼吸系统疾病提供了潜在的治疗目标与思路，不过，还需要进行更多的临床研究。总之，这些研究强调了 SCFAs 在减轻哮喘炎症方面的潜在作用。

除上述积极效应外，一些研究表明 SCFAs 可能对于感染性呼吸道疾病、结核病等也能发挥保护作用。丁酸盐、丙酸盐和乙酸盐这 3 种主要的 SCFAs 可有效预防呼吸道感染。根据 Haak 等（2018）的研究，接受异体造血干细胞移植的患者粪便中含有大量产生丁酸盐的细菌时，其患病毒性下呼吸道感染的概率较低。丁酸盐的保护作用不仅限于病毒，它还能有效延长感染肺炎克雷伯氏菌的小鼠的寿命（Chakraborty 等，2017）。此外，乙酸盐还能防止小鼠感染呼吸道合胞病毒，减少肺部病毒负荷和炎症。另一份报告也表明，乙酸盐通过激活 FFAR2 降低了细菌负荷和炎症水平，从而防止肺炎克雷伯氏菌感染（Antunes 等，2019）。研究还发现，乙酸盐通过增强肺泡巨噬细胞对抗细菌的能力，从而延长小鼠的存活时间，保护小鼠免受后续肺部感染（Galvão 等，2018）。此外，研究还表明，丁酸盐能激活 GPR109A，刺激 Treg 细胞以及产生 IL–10 和 IL–18 的细胞分化（Singh 等，2014）。在最近的一项研究中，流感发生时肠道微生物群的紊乱导致 SCFAs 乙酸酯的产生减少，这影响了在流感期间肺泡巨噬细胞杀灭造成超级感染的细菌的能力（Sencio 等，2020）。丙酸盐和丁酸盐在内的所有这些化合物都会被运送到肺部，在肺部它们可能会减少 IL–17 的合成，抑制 Th1 免疫，所有这些都可能会影响结核杆菌感染的发展。

值得注意的是，SCFAs 同时具有抗炎和促炎功能。体内和体外试验结果证实，SCFAs 可通过直接调节 T 细胞和树突状细胞来抑制 Th2 反应，从而改善肠道菌群失调驱动的肺部炎症（Cait 等，2018）。然而，SCFAs 通过 p38 MAPK 信号通路在原代人肺成纤维细胞和气道平滑肌细胞中显示出促炎症作用（Rutting 等，2018）（表 3.4）。

因此，SCFAs 可以降低慢性阻塞性肺病和急性呼吸衰竭的严重程度和发病率。这些作用主要是通过调节肺免疫稳态、气道上皮屏障和肠道微生物群来实现的。与造模后补充 SCFAs 相比，更多的研究关注肺病造模前补充 SCFAs，其预防效果明显。

表 3.4　SCFAs 对肺健康的作用研究

SCFAs 类型	模型 / 受试者	剂量和持续时间	结果	影响类型	参考文献
丁酸	成年雌性小鼠以低纤维食物喂养 4 周后接受丁酸盐	丁酸钠（500 mmol/L），持续 2 周	平衡先天性免疫和适应性免疫；↓流感病毒感染；↓免疫相关病理学	有益效果	Trompette 等，2018
乙酸 / 丙酸 / 丁酸	分化人支气管上皮细胞（16HBE）	10 mmol/L 乙酸酯、0.5 mmol/L 丙酸酯或 1 mmol/L 丁酸酯	↑肺免疫防御作用；↑气道上皮屏障中 TJ 蛋白的表达	有益效果	Richards 等，2020

续表

SCFAs 类型	模型 / 受试者	剂量和持续时间	结果	影响类型	参考文献
短链脂肪酸	雌性 C57BL/6 小鼠吸烟暴露肺气肿模型	柑橘来源的高纤维饮食	↓肺气肿发展；↓局部和全身炎症	有益效果	Jang 等，2021
乙酸 / 丙酸	用不同纤维含量的食物喂养 C57BL/6 雌性小鼠	乙酸钠或丙酸钠（200 mmol/L），持续 3 周	影响肺部的免疫环境；↓过敏性炎症的严重程度	有益效果	Trompette 等，2014
乙酸 / 丙酸 / 丁酸	万古霉素处理过的小鼠	40 mmol/L 丁酸盐、67.5 mmol/L 醋酸盐加 25.9 mmol/L 丙酸盐	↓ Th2 反应；↓肺部炎症	有益效果	Cait 等，2018
乙酸 / 丙酸 / 丁酸	原代人肺成纤细胞（HLFs）和气道平滑肌（ASM）细胞	乙酸盐（0.5 ～ 25 mmol/L）、丙酸盐（0.5 ～ 25 mmol/L）、丁酸盐（0.01 ～ 10 mmol/L）	↑ TNF α 诱导的炎症反应；通过激活 FFAR3 和 p38 MAPK 信号释放 IL-6 和 CXCL8	负面效果	Rutting 等，2019

3.6 短链脂肪酸对骨骼健康的影响

骨骼肌是人类最大的器官，在全身能量代谢中起着关键作用，SCFAs 通过与特异性受体 FFAR3 和 GPR109A 结合发挥作用。FFAR3 在结肠平滑肌中表达，SCFAs 以 FFAR3 依赖性方式诱导肌肉的阶段性收缩，改善骨骼功能（Tazoe 等，2009）。GPR109A 在破骨前体巨噬细胞中高度表达，在破骨细胞分化和骨吸收过程中发挥重要作用。Lucas 等（2018）通过饮水给小鼠补充 150 mmol/L 乙酸钠，为期 8 周，结果发现乙酸钠可以改善小鼠的骨形成。SCFAs 能够增加骨骼肌组织中的 AMP 表达量和 AMP/ATP 比率，从而诱导肌管和骨骼肌中 AMPK 的激活，进而诱导脂肪酸的摄取和氧化、葡萄糖的摄取和糖异生的增加，抑制脂肪生成和糖酵解（Mihaylova 等，2011）。SCFAs 还能够通过增加 AMPK 和 PPARγ 共激活因子 1α（PPAR-γ coactivator 1 alpha，pGC1α）的磷酸化影响骨骼肌葡萄糖和脂质代谢，增加胰岛素受体底物 1 的表达和蛋白激酶 b 的磷酸化，从而保持骨骼肌胰岛素的敏感性（Frampton 等，2020）。此外，SCFAs 是体内破骨细胞代谢和骨量的调节剂，用其处理可显著增加小鼠骨量并防止绝经后以及炎症引起的骨质流失；SCFAs 对骨量的保护作用与破骨细胞分化和骨吸收的抑制有关，包括下调破骨细胞基因 *TRAF6* 和 *NFATc1* 等。因此，SCFAs 作为破骨细胞代谢和骨量的调节剂，与特异性受体结合，AMPK 的激活可能是 SCFAs 改变骨骼肌代谢的关键机制。

肠道微生物群和 micro-CT 分析结果表明，马铃薯淀粉对骨量的积极影响可能与盲肠中更高比例的 SCFAs 的产生有关，这导致肠道和骨髓中促炎症基因表达减少，从而抑制细胞因子介导的破骨细胞的骨吸收（Zhang 等，2022）。根据研究报道，SCFAs 可通过维持骨平衡和骨骼肌功能来保护骨骼。

骨平衡的两个主要组成部分——破骨细胞和成骨细胞——分别对骨的吸收和形成起作用。骨密度降低和骨平衡失调是类风湿性关节炎等慢性炎症性疾病的特征，这是由于破骨细胞激活导致骨吸收增加所致。有研究显示，SCFAs，尤其是丙酸盐和丁酸盐，通过抑制 RANKL 信号传导阻止骨髓中破骨细胞前体细胞的发育，而不影响成骨细胞（Lucas 等，2018）。此外，研究表明丁酸盐抑制 HDAC 及其下游基因，从而抑制破骨细胞的形成（Rahman 等，2003）。

包括类风湿性关节炎、骨关节炎和痛风在内的全身性自身免疫性疾病的特点是骨和软骨的进行性损伤以及慢性关节炎症。最近的研究表明，纤维摄入量的增加已被证实对缓解痛风的发展具有积极作用，这可以通过产生 SCFAs 来解释。有研究表明，高纤维饮食和乙酸盐的摄入减轻了单钠尿酸盐结晶诱发的痛风相关炎症。这种关系与乙酸盐引发中性粒细胞凋亡的能力有关（Vieira et al.，2017）。

关于丁酸盐给药对骨骼肌的有利影响，研究发现，对 C57BL/6J 小鼠补充丁酸盐可以防止高脂饮食引起的骨骼肌不完全氧化（Henagan 等，2015）。此外，丁酸盐通过 PI3K/Akt/mTOR 信号通路减少糖尿病肾病诱导的骨骼肌萎缩（Tang 等，2021）。因此，通过提高骨骼肌氧化能力，丁酸盐可部分预防肥胖和胰岛素抵抗（表 3.5）。

总之，SCFAs 通过调节肠道微生物群以及破骨细胞和成骨细胞之间的骨平衡来调节骨功能。SCFAs 对骨功能的作用包括降低类风湿性关节炎、骨关节炎、痛风和骨骼肌萎缩的严重程度。除了研究人员研究最深入的丁酸盐外，体内研究证明醋酸盐对宿主骨骼健康有益，对骨相关疾病具有预防和治疗作用。

表 3.5　SCFAs 对骨骼健康的作用研究

SCFAs 类型	模型 / 受试者	剂量和持续时间	结果	影响类型	参考文献
短链脂肪酸	1 日龄雄鸭	SCFAs（67.5 mmol/L 乙酸酯、38.8 mmol/L 丙酸酯、22.8 mmol/L 丁酸酯），持续 14 天	↓ 肠道和骨髓中的促炎基因表达；↓ 破骨细胞骨吸收	有益效果	Zhang 等，2022
短链脂肪酸	C57BL/6J 雌性小鼠、卵巢切除小鼠、关节炎模型	150 mmol/L 乙酸盐、丙酸盐和丁酸盐，持续 8 周	↑ 骨量；↓ 绝经后和炎症引起的骨质流失；↓ 破骨细胞分化和骨吸收	有益效果	Lucas 等，2018
丁酸	大鼠骨髓细胞和 RAW-D 细胞	丁酸盐（1 mmol/L）24 h	↓ 破骨细胞特异性信号；↓ HDAC 活性调节破骨细胞生成过程	有益效果	Rahman 等，2003
乙酸	单钠尿酸（MSU）诱导的雄性 C57Bl/6 小鼠痛风模型	在饮用水中添加 150 mmol/L 乙酸盐，口服丁酸盐（50 mmol/L）或丙酸盐（25 mmol/L）	↑ caspase 依赖性中性粒细胞凋亡、流出；↓ 炎症反应	有益效果	Vieira 等，2017
丁酸	饮食性肥胖 C57BL/6J 小鼠	丁酸钠，5 g/（kg · 天），按正常每日能量摄入量计算	↓ 饮食诱导的小鼠胰岛素抵抗；↑ 能量消耗；↑ 线粒体功能	有益效果	Gao 等，2009
丁酸	5 周龄雄性 C57BL/6J 小鼠	5% 丁酸盐（重量比），持续 10 周	↑ 胰岛素增敏和抗肥胖；↑ 肌肉线粒体功能	有益效果	Henagan 等，2015
乙酸	6 周龄 C57Bl/6 小鼠	100 mmol/L、200 mmol/L 或 300 mmol/L 乙酸钠，持续 3 周	↓ 结肠炎严重程度	有益效果	Macia 等，2015

续表

SCFAs类型	模型 / 受试者	剂量和持续时间	结果	影响类型	参考文献
丁酸	4 周龄雄性 db/db 和 db/m 小鼠	以正常的每日热量摄入量为标准，每天 1 g/kg，持续 12 周	↑糖尿病、肾病诱发的肌肉萎缩缓解效果	有益效果	Tang 等，2022

3.7 短链脂肪酸对心血管的影响

现有证据表明，SCFAs 可被心脏快速吸收和氧化，为心脏提供能量并预防心力衰竭（Palm 等，2022）。例如，由于微生物群多样性减少、产生丁酸盐的菌株较少，以及 SCFAs 主要由微生物在体内产生等因素，心衰患者产生 SCFAs 的能力总体有限（Jin 等，2020）。一项研究显示，虽然 SCFAs 对心脏 ATP 合成的贡献相对较低，但心衰患者中的乙酸提取量增加了约 20%。这一结果表明，增加循环系统中的 SCFAs 水平可能有利于心衰患者的能量生成（Murashige 等，2020）。有报道称，在健康和衰竭的心脏中，丁酸盐的 ATP 合成明显高于酮体，这表明 SCFAs 是更有效的能量生成物质（Carley 等，2021）。

高血压被认为是心血管疾病的主要危险因素，它受到遗传和环境的影响，其高发病率使其成为全球主要的健康挑战。研究发现，膳食纤维在肠道微生物群厌氧发酵作用下产生的 SCFAs 参与了血压的调节，即当肠道中的 SCFAs 浓度较低时，血压也会受到影响（Yang 等，2020）。醋酸盐补充剂已被证实对减少阻塞性睡眠呼吸暂停和高血压小鼠模型的高血压发展具有有益作用（Poll 等，2021）。丙酸盐干预可能通过调节 Treg 细胞减少全身炎症来降低血压（Bartolomaeus 等，2019）。血管组织中的 GPR 受体以及嗅觉受体 78 作为丙酸盐降低血压机制的可能性已被发现（Natarajan 等，2016）。研究发现，外源性 SCFAs 可降低血压，而 GPR41/43 抑制剂可减轻这些影响（Onyszkiewicz 等，2020）。GPR41/43 受体和结肠迷走神经信号参与了丁酸盐抑制高血压发病的机制（Onyszkiewicz 等，2019）。与野生型大鼠相比，对盐敏感的 Gper1–/– 大鼠血压更低，肠道疾病更少，证明作为 GPR 受体之一的 GPER1 与血压调节有关（Waghulde 等，2018）。作为一种双向调节因子，高血压还可能通过降低肠道微生物群的丰度和多样性来破坏肠道微生物群结构，从而降低 SCFAs 的浓度（Yang 等，2020）。

目前，动脉粥样硬化被认为是一种发生在大动脉的慢性炎症性疾病。不同的 SCFAs 在动脉粥样硬化中具有不同的功能（Yao 等，2022）。根据研究，服用丙酸盐补充剂可通过调节肠道免疫系统和血管炎症缓解动脉粥样硬化（Haghikia 等，2021）（表 3.6）。

综上，SCFAs 可影响能量产生和血管炎症，进而影响心血管功能。这些结果的有益影响包括降低心力衰竭、高血压和动脉粥样硬化的发病率。

表 3.6　SCFAs 对心血管健康的作用研究

SCFAs 类型	模型 / 受试者	剂量和持续时间	结果	影响类型	参考文献
游离脂肪酸	110 名患有或未患有心力衰竭的受试者	人群血浆代谢物	乙酸盐的摄取量与循环中乙酸盐的浓度成正比	有益效果	Murashige 等，2020
丁酸	TAC 诱导心脏肥大 SD 大鼠，假手术为对照组	0.5 mmol/L 丁酸盐和 0.5 mmol/L 3- 羟基丁酸盐混合液	↑衰竭心脏的氧化作用	有益效果	Carley 等，2021
丙酸	野生型 NMRI 和脂蛋白 E 基因敲除缺陷小鼠	200 mmol/L	↓心脏肥大、纤维化、血管功能障碍和高血压	有益效果	Bartolomaeus 等，2019
①短链脂肪酸；②乙酸钠	①雌雄 C57BL/6J 小鼠；②雄性 C57BL/6J 小鼠	① 1 g/kg ② 1 mol/L	① ↓ MAP、HR 和心脏收缩力；② ↓长时间暴露后的心率	有益效果	Poll 等，2021
乙酸、丙酸	① C57BL/6 Gpr41 WT 雄性小鼠；② Gpr41 KO 小鼠	乙酸 0.3 ～ 10 mmol/L 丙酸 0.3 ～ 10 mmol/L	↓低血压反应	有益效果	Natarajan 等，2016

续表

SCFAs 类型	模型 / 受试者	剂量和持续时间	结果	影响类型	参考文献
丁酸	14～16 周大的 Wistar 大鼠	1.4 mmol/kg、2.8 mmol/kg 和 5.6 mmol/kg	↑降压作用，似乎由结肠传入神经信号和 GPR41/43 受体介导	有益效果	Onyszkiewicz 等，2019
丙酸	① 16 周雌性 C57BL/6、载脂蛋白 $E^{-/-}$ 小鼠 ② 16 周雌性 C57BL/6、载脂蛋白 $E^{-/-}$ 小鼠	① 150 mmol/L 丙酸钙，4 周 ②丙酸钙（500 mg），每天 2 次，共 8 周	①↓高脂饮食诱发载脂蛋白 $E^{-/-}$（$Apoe^{-/-}$）小鼠高胆固醇血症和动脉粥样硬化 ②↓高胆固醇血症小鼠的血清低密度脂蛋白和总胆固醇	有益效果	Haghikia 等，2022

总之，人类不仅可以通过肠道微生物群对非消化性碳水化合物的内源性生物合成获得 SCFAs，还可以直接通过食物获得 SCFAs。此外，人体健康，包括肠、脑、肾、肝、肺、骨骼等器官的功能，都受到内源性合成和外源性摄入 SCFAs 的显著影响。

SCFAs 对人体健康的调节具有双面性，既有积极作用，也有消极作用，但 SCFAs 对人体的积极作用更为明显。

目前的研究主要以器官为关注点分析了 SCFAs 的功能，但这可能并不能准确全面地代表人体的整体健康。例如，我们尚不了解 SCFAs 作用的器官之间是否可以发挥相互作用。只有充分理解 SCFAs 如何影响机体健康，才能对如何预防和控制疾病的发生有新的认识。

3.8 短链脂肪酸在疾病改善中的应用

Tamang 等（2016）指出肥胖是心血管疾病、血脂异常、肝胆疾病、糖尿病、过早死亡和几种癌症的危险因素。据估计，世界上有 17 亿人体重超标。SCFAs 可以与肠细胞表面受体 GPR41/43 作用，促进肠道激素分泌，这些激素可以调节胃肠道运动和分泌，改变肠壁通透性，减少胃肠道从食物中吸收能量。此外，超重儿童粪便中 SCFAs 的浓度（特别是丁酸和丙酸）高于正常体重儿童。在 Polyviou 等（2016）进行的一项随机对照交叉研究中，发现以菊粉为对照，菊粉丙酸酯含 27wt% 的丙酸盐能明显减少食物摄入量，表明结肠丙酸盐在食欲调节中发挥作用。

SCFAs 在精神疾病中也发挥着其独特作用，对于初治精神分裂症患者，血清丁酸盐水平升高与良好的治疗反应相关（Li 等，2021）。膳食 SCFAs 通过抑制 JNK1 和 p38 通路，帮助扩张肠道 Tregs 以调节自身免疫应答，从而减少自身免疫性脑脊髓炎患者的轴索损伤（Haghikia 等，2015）。将精神分裂症患者的粪便微生物群移植到接受抗生素治疗的小鼠中可导致受体动物的异常行为，例如精神运动功能亢进以及学习和记忆受损，与接受健康对照的粪便相比，这些小鼠还显示出外周和大脑中色氨酸降解的犬尿氨酸 – 犬尿酸途径的升高，以及前额叶皮层中的基础细胞外多巴胺和海马体中的 5– 羟色胺增加（Zhu 等，2020）。另一项研究显示，小鼠在经过 3 周的社交失败和拥挤实验（心理学的一个经典实验，可能易使人变得不安，过度拥挤可能导致行为失常）后，给予小鼠醋酸盐、丙酸盐和丁酸盐的混合物，可缓解应激反应性增强和应激引起的肠道通透性增加，同时在露天试验中减少焦虑样行为，在强制游泳试验中减少抑郁样行为（Kelly 等，2021）。Ohara（2019）研究表明，乳酸菌、双歧杆菌和梭状芽孢杆菌等产生的一些 SCFAs 活性物质，如 γ – 氨基丁酸（Gaba）在各种精神疾病中具有非常特殊的功能，可能是潜在的新型精神生物制剂。

SCFAs 已被证明具有抗癌潜力，丁酸在其中占有重要地位。丁酸是厚壁菌门的主要代谢产物，能被结肠上皮细胞吸收利用，是结肠、盲肠能量的首选来源（Susan，2002）。比利时农业部动物营养与饲养部门曾将丁酸盐添加到仔猪日粮中，结果发现，与对照组相比，日粮中添加丁酸盐可使仔猪日增重 17%，提高采食量 7.3%，对一些性能参数有一定的改善作用。另外，丁酸还能抑制肿瘤细胞的增殖分化、调节基因表达、维持肠道内环境稳定，对结肠炎和结肠癌起到预防作用（Kotunia 等，2004）。傅红等（2003）以人结肠癌 Caco–2 细胞为模型，研究了 3 种短链脂肪酸对细胞增殖、分化和转移的影响。结果表明 3 种短链脂肪酸，尤其是丁酸盐，可影响人结肠癌细胞的表型，显著延长肿瘤细胞倍增时间，增强癌细胞分化标志物蛋白酶的表达，并明显抑制癌细胞的转移。试验结果还表明，高丁酸盐 / 醋酸盐比率的纤维素饮食对抗结肠癌作用起到积极的效果。

Trompette 等（2014）发现喂食高纤维饮食的小鼠表现出 SCFAs 水平升高，肺部过敏性炎症易感性降低，而喂食低纤维饮食的小鼠则显示 SCFAs 水平降低，过敏性气道疾病增加。膳食纤维可改变肠道 SCFAs 水平，维持黏膜稳态和肠上皮完整性，促进 Tregs 的生长，并抑制炎性细胞因子的表达，以预防和 / 或改善疾病（Tan，2023）。

4 短链脂肪酸的应用

短链脂肪酸（SCFAs）的广泛生物学功能决定了其具有多重应用场景，如婴幼儿配方乳粉和功能性食品和食品工业等方面。

4.1 短链脂肪酸在婴幼儿配方乳粉中的应用

丁酸是一种短链饱和脂肪酸，其最大功能是供应能量，而聚集在脂细胞内形成脂肪的较少，具有参与基因调控、调节免疫应答和炎症反应，抑制肿瘤生长，促进细胞分化和凋亡的作用。己酸、辛酸、癸酸 3 种饱和脂肪酸具有相似的生物活性，例如，辛酸和癸酸具有抗病毒的生物活性，癸酸和甘油三酯生成的癸酸单酰甘油酯具有抗艾滋病毒的功能（李晓敏等，2021）。乳粉中天然含有丁酸、己酸、辛酸和癸酸，虽然不属于婴幼儿配方乳粉中常规审核及检测项目，但其对婴幼儿生长发育同样起到了一定的积极作用（崔承远，2021）。

戴昕悦等（2020）对人乳短链脂肪酸与配方乳粉中的短链脂肪酸进行比较发现，不同配方乳粉中的短链脂肪酸含量差异较大，但总体含量比人乳中的短链脂肪酸含量高。此外，有研究人员分析了目前中国市场上婴幼儿配方乳粉脂肪酸的组成，并且比较了饱和脂肪酸和不饱和脂肪酸的质量分数差异（表 4.1）。短链饱和脂肪酸可以直接经门静脉进入肝脏，迅速分解转换为能量，更容易被机体吸收。乳脂配方乳粉组的短链饱和脂肪酸质量分数明显高于植物油配方乳粉组，植物油几乎不含短链饱和脂肪酸，牛乳中含有一定程度的短链饱和脂肪酸，质量分数约为 6.12%（Ceballos 等，2009）。因此，短链饱和脂肪酸水平可能是以乳脂为基础的婴幼儿配方乳粉的标志。

表 4.1 市售婴幼儿配方乳粉的饱和脂肪酸组成（李琳瑶等，2022）

脂肪酸名称	植物油配方乳粉组（%）	乳脂配方乳粉组（%）
丁酸（C4:0）	ND	ND
己酸（C6:0）	0.15±0.04	0.67±0.28
辛酸（C8:0）	1.58±0.45	0.69±0.40
癸酸（C10:0）	1.26±0.32	1.18±0.54

续表

脂肪酸名称	植物油配方奶粉组（%）	乳脂配方奶粉组（%）
十一碳酸（C11:0）	ND	ND
月桂酸（C12:0）	9.67±1.98	3.18±1.94
十三碳酸（C13:0）	ND	ND
肉豆蔻酸（C14:0）	4.06±0.73	4.26±1.84
十五碳酸（C15:0）	0.05±0.03	0.36±0.13
棕榈酸（C16:0）	17.12±6.54	21.77±5.57
十七碳酸（C17:0）	0.07±0.02	0.24±0.08
硬脂酸（C18:0）	3.69±0.44	5.51±0.78
花生酸（C20:0）	0.30±0.03	0.28±0.04
二十一碳酸（C21:0）	ND	ND
山嵛酸（C22:0）	0.38±0.12	0.31±0.11
二十三碳酸（C23:0）	ND	ND
二十四碳酸（C24:0）	0.18±0.05	0.18±0.08
饱和脂肪酸	38.51±5.02	38.62±8.06
短链饱和脂肪酸	0.15±0.04	0.67±0.28
中链饱和脂肪酸	12.52±2.64	5.04±2.82
长链饱和脂肪酸	25.85±6.34	32.90±6.43

因此，婴儿配方乳粉在配方设计时，应以母乳为参考，同时结合食品安全国家标准和婴儿的营养需求，保证其合理性。在脂肪酸添加方面，需考虑各种脂肪酸的组成、质量分数，特别是具有特殊意义的脂肪酸的强化和合理搭配，以及脂肪酸在甘油三酯的位置分布，以更好地为婴儿的生长发育提供有力的支撑，研制出真正适合中国宝宝的配方乳粉。

4.2 短链脂肪酸在功能性食品中的应用

肠道微生物群发酵产生的 SCFAs 与积极的新陈代谢效应有关。Joseph（2019）在一项调查中探究了 40 名学龄儿童（19 名正常体重儿童和 21 名体重超标儿童）饮用益生菌后的肠道菌群变化。受试者被分为正常体重组和体重超标组，并进一步分为干预组和对照组，这些儿童饮用含有干酪乳杆菌 Shirota 菌株的益生菌饮料 4 周。研究发现，饮用益生菌饮料导致体重超标儿童肠道菌群中乳酸菌和双歧杆菌数量显著增加。在正常体重和体重超标的儿童补充 4 周后，观察到两组的 SCFAs 浓度都显著增加，特别是丙酸。此外，体重超标儿童粪便中总 SCFAs 的浓度增加，特别是丁酸和丙酸。

Cheng 等（2022）发现 Gaba 是一种四碳短链脂肪酸，具有多种生理活性，包括促进神经元发育、缓解焦虑和失眠、降压、降血糖、抗肿瘤、抗炎、抗菌、抗过敏、保护肝脏、肾脏和肠道等，是一种卓越的健康效应因子。早在 2009 年，Gaba 就已被我国卫生部列为新资源食品。目前 Gaba 已实现产业化，并广泛用于各类食品和保健品中。

养成健康的饮食习惯比药物、外科和保健品更可行，具有一定的优势。为了促进 SCFAs 更好地发挥作用，研究人员评估了提高膳食来源 SCFAs 的可行性。对常用食品和饮料中的 SCFAs 进行量化分析，发现在乳制品中奶酪含有相对较高的丙酸盐和丁酸盐含量（表 4.2）。

表 4.2　不同乳制品中短链脂肪酸（SCFAs）的含量（Gill 等，2020）

食物	份量（g）（体积）	乙酸盐		丙酸盐		丁酸盐	
		mg/g	mg/ 份	mg/g	mg/ 份	mg/g	mg/ 份
切达干酪	30	0.90	26.98	0.00	0.00	0.22	6.61
蓝纹奶酪	25	1.32	32.93	0.00	0.00	5.46	136.54
布里奶酪	25	0.33	8.30	0.00	0.00	4.24	105.91
瑞士奶酪	30	0.80	23.97	3.64	109.2	0.11	3.42
全脂牛奶	258（1 杯，250 mL）	0.01	3.35	0.00	0.00	0.00	0.00
低脂牛奶	258（1 杯，250 mL）	0.02	4.13	0.00	0.00	0.02	4.90
希腊风味酸奶	100	0.15	15.19	0.00	0.00	0.03	3.28
酸奶油	30	0.47	14.16	0.00	0.00	0.02	0.66
黄油	5	0.03	0.17	0.00	0.00	0.01	0.05
开菲尔	206（200 mL）	0.48	98.87	0.00	0.00	0.02	3.10

Annunziata 等（2020）提出可以使用不同的微生物进行发酵，以生产富含短链脂肪酸的功能性食品和饮料，从而提高食品和饮料的营养价值（图 4.1）。该想法表明 SCFAs 在功能性食品的研发过程中仍然充满巨大前景，其既不会对环境产生额外影响，也能够满足特定人群的营养需求。基于此，我们更需要进一步理解 SCFAs 的功能，为其应用于食品和医药保健品奠定更加完善的理论基础。同时，还需要正视 SCFAs 在功能性食品中可能发挥的潜力和作用，通过干预饮食或保健品对与 SCFAs 相关的疾病进行调控具有长远意义。

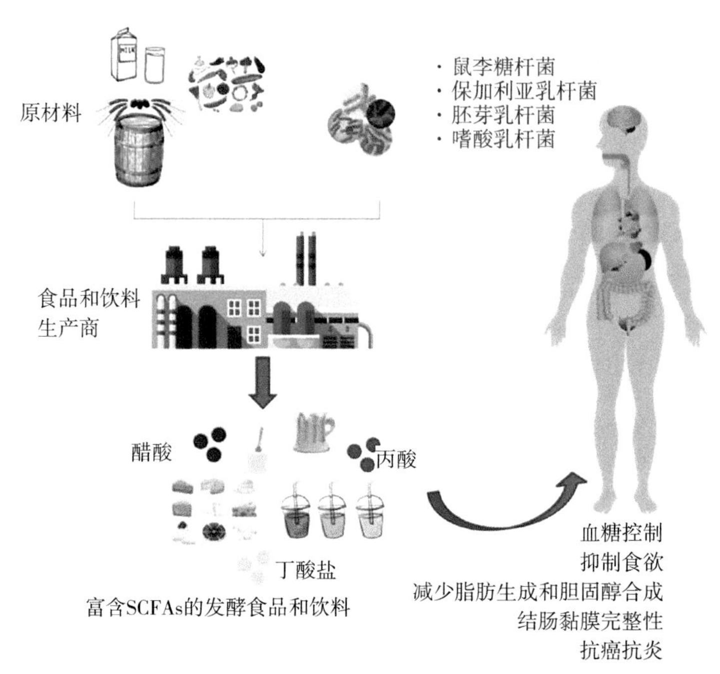

图 4.1　富含 SCFAs 的功能性食品和饮料对人类健康产生的多种有益影响（Annunziata，2020）

4.4　短链脂肪酸在食品工业中的应用

除用作功能性食品外，SCFAs 在食品工业中也有重要作用。由于短链脂肪酸的抗菌特性，食品制造商和加工商更经常使用它们来确保食品安全（图 4.2）。

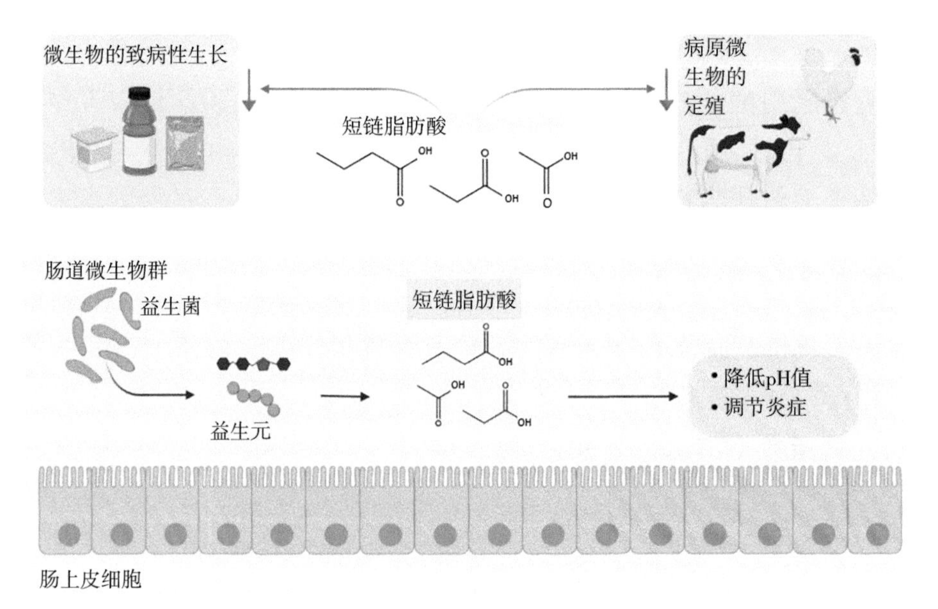

图 4.2 SCFAs 在食品工业中的适用性，降低病原菌感染的可能性

SCFAs 被加入各种食品中，以抑制微生物的致病性生长（Ricke，2003）。因此，在牲畜饲料中补充 SCFAs 已证明有可能避免病原微生物的脱落和定植，从而最大限度地降低感染疾病的风险（Van Immerseel 等，2003）。据报道，短链脂肪酸和肠道微生物群对肉类生产具有有益影响，因此可用于肉兔行业（Fang 等，2020）。丙酸钠可以作为抗菌剂用于肉类制剂、肉类加工品和鱼类加工品。根据欧盟 1333/2008 号法规附件Ⅱ，丙酸钠目前是欧盟授权的食品添加剂，可用于几类食品（烘焙食品和奶酪产品），最高含量为 3000 mg/kg（表 4.3）。2014 年，欧洲食品安全局食品添加剂和食品营养源（ANS）专家小组对丙酸（E 280）和丙酸盐（E 281–283）作为食品添加剂的安全性进行了重新评估，未发现对消费者的安全问题。在对犬进行的一项为期 90 d 的研究中，引起接触部位效应的丙酸浓度是目前食品中丙酸最高允许浓度的 3 倍。而在一项为期 104 周的研究中，饮食中单一浓度为 5.1% 的丙酸钠无负面影响报告，这表明丙酸钠的刺激性比丙酸本身小，其丙酸钠浓度比肉类制剂、加工肉类和鱼类中可能出现的拟议最高允许含量的食物浓度高约 10 倍。因此，丙酸钠在降低刺激性方面也许优于丙酸（表 4.4）。该研究建议将丙酸钠（E 281）作为食品添加剂在肉类制剂、加工肉类和鱼类中的使用范围扩大至每千克 5000 mg，

该剂量不会引起安全问题（ANS，2016）。

短链脂肪酸改性淀粉皮克林（Pickering）乳液的表征和稳定性会更好，一项研究采用不同改性程度和淀粉浓度的乙酰化、丙酰化和丁酰化大米和藜麦淀粉作为稳定油分含量为 10% 的水包油淀粉皮克林乳剂。对乳化后储存 50 d 后的短链脂肪酸改性淀粉皮克林乳剂（SPEs）进行粒度分布、微观结构、乳化指数和稳定性的评估，发现淀粉浓度的增加导致乳液液滴尺寸减小。通过增加 SCFAs 的链长可提高 SPEs 的乳化能力。链长较长（即丙酰化和丁酰化）的改性藜麦淀粉在较高的改性水平下显示出较高的乳化指数（>71%）和整个 50 d 储存期间的稳定性。通过优化配方，SCFAs 淀粉颗粒具有稳定功能食品、药物配方或工业食品应用中的乳液的潜力（Abdul，2020）。

表 4.3　丙酸 – 丙酸盐（E280–283）在食品中的最大残留限量（ANS，2016）

类别编号	食品	限制	最高水平（mg/L 或 mg/kg，视情况而定）
01.7.2	成熟奶酪	仅表面处理	适量
01.7.6	奶酪制品（不包括属于类别 16 的商品）	只有成熟的产品表面处理	适量
01.8	乳制品类似物，包括饮料增白剂	仅奶酪类似物（仅限表面处理）	适量

注：根据经修订的欧盟 No 1333/2008 法规附件Ⅱ所得的数据。

表 4.4　丙酸钠（E 281）用作食品添加剂时，按现行最大允许摄入量和拟议的扩展用途及使用量计算的估计暴露量（ANS，2016）

项目	幼儿（12~35 个月）	孩子（3~9 岁）	青少年（10~17 岁）	成年人（18~64 岁）	老年人（≥ 65 岁）
当前最高允许摄入量					
平均值（mg/kg）	0.61~22.99	0.61~22.99	0.61~22.99	0.61~22.99	0.61~22.99
最高水平（mg/kg）	3.18~39.26	4.03~43.53	12.31~22.07	0.74~15.42	0.66~13.60
当前最高允许摄入量和拟议的最大扩展用途					
平均值（mg/kg）	2.70~32.6	9.39~29.56	8.17~19.18	2.91~13.35	2.28~10.56
最高水平（mg/kg）	22.07~46.72	26.25~49.35	13.98~33.09	10.40~24.66	9.86~17.11

微胶囊化是一种将液体或固体颗粒的小液滴涂上一层保护性生物材料薄膜从而防止热和光损伤，并与其他食品成分分离的技术，SCFAs 在其中可扮演重要角色。在含有 SCFAs 的发酵渗透液样品中分别加入 5% 的麦芽糊精和 5% 的阿拉伯胶的混合物，然后进行喷雾干燥，发现其混合物表面光滑，大小分布均匀。由于具有流动性，相应的微胶囊被认为是食品工业用途的最佳选择。此外，由于这些胶囊的直径分布均匀，可使食品的风味分布均匀（Teixeira 等，2004）。

SCFAs 的抑菌、抗炎、抗肿瘤、免疫调节等生物活性近些年来已经得到了诸多研究人员的重视和证实，且其作为一种天然无毒无害的产物，在未来被广泛运用于功能性食品等领域具有极大的潜在发展可能性。

参考文献

崔承远, 2021. 婴幼儿配方奶粉中丁酸、己酸、辛酸、癸酸的测定方法验证 [J]. 中国乳业 (12):105–110.

戴昕悦, 袁婷兰, 金青哲, 等, 2020. 人乳短链脂肪酸的组成、检测方法与消化代谢研究进展 [J]. 中国油脂, 45(1): 27–30.

丁岩, 王娟, 张迪, 2019. 气相色谱法测定发酵乳中的 7 种短链脂肪酸 [J]. 食品与发酵工业, 45(2): 202–206.

董霞, 朱研, 沈林園, 等, 2022. 短链脂肪酸对肠上皮紧密连接的作用机制研究进展 [J]. 动物营养学报, 34(11):6936–6942.

李琳瑶, 华家才, 康巧娟, 等, 2022. 婴儿配方奶粉中脂肪酸的研究 [J]. 中国乳品工业, 50(6): 39–43.

林杨凡, 林炫财, 孔晶晶, 等, 2022. 短链脂肪酸调控肠道健康研究进展 [J]. 现代消化及介入诊疗, 27(4):520–524.

武旭芳, 张养东, 郑楠, 等, 2022. 乳中短链脂肪酸组成、生理功能及检测技术研究进展 [J]. 动物营养学报, 34(4): 2148–2155.

杨雪, 高亚男, 王加启, 2023. 短链脂肪酸在肠脑轴中的作用 [J]. 动物营养学报, 35(3): 1368–1379.

杨雪, 高亚男, 王加启, 等, 2023. 短链脂肪酸的功能研究进展 [J]. 食品科学, 44(13):408–417.

ABDUL HADI N, MAREFATI A, MATOS M, et al., 2020. Characterization and stability of short–chain fatty acids modified starch Pickering emulsions[J]. Carbohydr Polym., 240:116264.

AGUS A, CLEMENT K, SOKOL H, 2021. Gut microbiota–derived metabolites as central regulators in metabolic disorders[J]. Gut, 70(6): 1174–1182.

ANDRADE–OLIVEIRA V, AMANO M T, CORREA–COSTA M, et al., 2015. Gut bacteria products prevent aki induced by ischemia–reperfusion[J]. Journal of the American Society of Nephrology, 26:1877–88.

ANNUNZIATA G, ARNONE A, CIAMPAGLIA R, et al., 2020. Fermentation of foods and beverages as a tool for increasing availability of bioactive compounds focus on short–chain fatty acids[J]. Foods, 9(8):999.

ANTUNES K H, FACHI J L, DE PAULA R, et al., 2019. Microbiota–derived acetate protects against respiratory syncytial virus infection through a GPR43–type 1 interferon response[J]. Nature Communications, 10 (1): 3273.

ARAúJO J R, TAZI A, BURLEN–DEFRANOUX O, et al., 2020. Fermentation products of commensal bacteria alter enterocyte lipid metabolism[J]. Cell Host & Microbe, 27 (3):358–375.

ATSUMI S, HANAI T, LIAO J C, 2008. Non–fermentative pathways for synthesis of branched–chain higher alcohols as biofuels[J]. Nature, 451(7174): 86–99.

AZQUEZ–LANDAVERDE P A, TORRES J A, QIAN M C, 2006. Effect of high–pressure–moderate–temperature processing on the volatile profile of milk[J]. Journal of Agricultural and Food Chemistry, 54: 9184–9192.

BARCELO A, CLAUSTRE J, MORO F, et al., 2000. Mucin secretion is modulated by luminal factors in the isolated vascularly perfused rat colon[J]. Gut, 46(2): 218–224.

BARTHOLOME A L, ALBIN D M, BAKER D H, et al., 2004.Supplementation of total parenteral nutrition with butyrate acutely increases structural aspects of intestinal adaptation after an 80% jejunoileal resection in neonatal piglets[J]. Journal of Parenteral and Enteral Nutrition, 28(4):210–222.

BARTOLOMAEUS H, BALOGH A, YAKOUB M, et al., 2019. Short–chain fatty acid propionate protects from hypertensive cardiovascular damage[J]. Circulation, 139(11):1407–1421.

BEHARY J, AMORIM N, JIANG X T, et al., 2021. Gut microbiota impact on the peripheral immune response in non–alcoholic fatty liver disease related hepatocellular carcinoma[J]. Nature Communications, 12(1):187.

BOLOGNINI D, TOBIN A B, MILLIGAN G, 2015. The pharmacology and function of short chain fatty acid receptors[J]. Mol.Pharmacol, 89: 388–398.

BONAZ B, BAZIN T, PELLISSIER S, 2018. The vagus nerve at the interface of the microbiota–gut–brain axis [J]. Frontiers in Neuroscience, 12(49): 00049.

BUDDEN K F, GELLATLY S L, WOOD D L A, et al., 2017. Emerging pathogenic links between microbiota and the gut–lung axis[J]. Nature Reviews Microbiology, 15 (1):55–63.

BYASS P, 2014. The global burden of liver disease: a challenge for methods and for public health[J]. BMC Medicine, 12 (1):159.

CAIT A, HUGHES M R, ANTIGNANO F, et al., 2018. Microbiome–driven allergic lung inflammation is ameliorated by short–chain fatty acids[J]. Mucosal Immunology, 11(3):785–795.

CANI P D, AMAR J, IGLESIAS M A, et al., 2007. Metabolic Endotoxemia Initiates Obesity and Insulin Resistance[J]. Diabetes, 56(7):1761–1772.

CARLEY A N, MAURYA S K, FASANO M, et al., 2021. Short–chain fatty acids outpace ketone oxidation in the failing heart[J]. Circulation, 143(18):1797–1808.

CHAKRABORTY K, RAUNDHAL M, CHEN B B, et al., 2017. The mito–DAMP cardiolipin blocks IL–10 production causing persistent inflammation during bacterial pneumonia[J]. Nature Communications, 8:13944.

CHAMEKH L, CAIVO M, KHORCHANI T, et al., 2020. Impact of management system and lactation stage on fatty acid composition of camel milk[J]. Journal of Food Composition and Analysis, 87: 103418.

CHANG A J, ORTEGA F E, RIEGLER J, et al., 2015. Oxygen regulation of breathing through an olfactory receptor activated by lactate[J]. Nature, 527: 240–244.

CHANG Y H, JEONG C H, CHENG W N, et al., 2021. Quality characteristics of yogurts fermented with short–

chain fatty acid-producing probiotics and their effects on mucin production and probiotic adhe-sion onto human colon epithelial cells[J]. Journal of Dairy Science, 104(7): 7415-7425.

CHEN R, WU P, CAI Z, et al., 2019. Puerariae Lobatae Radix with chuanxiong Rhizoma for treatment of cerebral ischemic stroke by remodeling gut microbiota to regulate the brain-gut barriers[J]. Journal of Nutritional Biochemistry, 65:101-114.

CHEN R, XU Y, WU P, et al., 2019. Transplantation of fecal microbiota rich in short chain fatty acids and butyric acid treat cerebral ischemic stroke by regulating gut microbiota[J]. Pharmacological Research, 148:104403.

CHENG Y, LIU J, LING Z, 2022. Short-chain fatty acids-producing probiotics:A novel source of psychobiotics[J]. Critical Reviews in Food Science and Nutrition, 62(28): 7929-7959.

CONTARINI G, PELIZZOLA V, SCURATI S, et al., 2017. Polar lipid of donkey milk fat: Phospholipid, ceramide and cholesterol composition[J]. Journal of Food Composition and Analysis, 57: 16-23.

CUMMINGS J H, POMARE E W, BRANCH W J, et al., 1987. Short chain fatty acids in human large intestine, portal, hepatic and venous blood[J]. Gut, 28(10):1221-1227.

CZYZAK-RUNOWSKA G, WOJTOWSKI J A, DANKOW R, et al., 2021. Mare's milk from a small polish specialized farm-basic chemical composition, fatty acid profile, and healthy lipid indices[J]. Animals, 11(6): 1590.

DAI X Y, YUAN T L, ZHANG X H, et al., 2020. Short-chain fatty acid (SCFA) and medium-chain fatty acid (MCFA) concentrations in human milk consumed by infants born at different gestational ages and the variations in concentration during lactation stages[J]. Food and Function, 11(2): 1869-1880.

DALILE B, VAN OUDENHOVE L, VERVLIET B, et al., 2019. The role of short-chain fatty acids in microbiota-gut-brain communication[J]. Nature Reviews Gastroenterology & Hepatology, 16(8):461-478.

DANG A T, MARSLAND B J, 2019. Microbes, metabolites, and the gut-lung axis[J]. Mucosal Immunology, 12 (4):843-850.

DANGANA E O, OMOLEKULO T E, AREOLA E D, et al., 2019. Sodium acetate protects against nico-tine-induced excess hepatic lipid in male rats by suppressing xanthine oxidase activity[J]. Chemico-Biological Interactions, 316:108929.

DE VADDER F, KOVATCHEVA-DATCHARY P, GONCALVES D, et al., 2014. Microbiota-generated metabolites promote metabolic benefits via gut-brain neural circuits[J]. Cell, 156(1-2): 84-96.

DE WIT N, DERRIEN M, BOSCH-VERMEULEN H, et al., 2012. Saturated fat stimulates obesity and hepatic steatosis and affects gut microbiota composition by an enhanced overflow of dietary fat to the distal intestine[J]. American Journal of Physiology-Gastrointestinal and Liver Physiology, 303(5):G589-G599.

DEB-CHOUDHURY S, BERMINGHAM E N , YOUNG W , et al., 2018. The effects of a wool hydrolysate on short-chain fatty acid production and fecal microbial composition in the domestic cat (Felis catus) [J]. Food Funct. 9(8):4107-4121.

DEN BESTEN G, BLEEKER A, GERDING A, et al., 2015. Short-chain fatty acids protect against high-fat diet-induced obesity via a PPAR γ -dependent switch from lipogenesis to fat oxidation[J]. Diabetes, 64(7):2398-

2408.

DIAO H, JIAO A R, YU B, et al., 2019. Gastric infusion of short-chain fatty acids can improve intestinal barrier function in weaned piglets[J]. Genes and Nutrition, 14: 4.

DINAN T G, CRYAN J F, 2017. Gut instincts: microbiota as a key regulator of brain development, ageing and neurodegeneration[J]. The Journal of Physiology, 595(2): 489–503.

D'SOUZA W N, DOUANGPANYA J, MU S, et al., 2017. Differing roles for short chain fatty acids and GPR43 agonism in the regulation of intestinal barrier function and immune responses[J]. PLoS One, 12(7): e0180190.

DUNCAN S H, BARCENILLA A, STEWART C S, et al., 2002. Acetate utilization and butyryl coenzyme A (CoA):acetate-CoA transferase in butyrate-producing bacteria from the human large intestine[J]. Applied and Environmental Microbiology, 68(10): 5186–5190.

DUPRAZ L, MAGNIEZ A, ROLHION N, et al., 2021. Gut microbiota-derived short-chain fatty acids regulate IL-17 production by mouse and human intestinal gammadelta T cells[J]. Cell Reports, 36(1): 109332.

EFSA ANS Panel (EFSA Panel on Food Additives and Nutrient Sources added to Food), 2016. Scientific opinion on the safety of the extension of use of sodium propionate (E 281) as a food additive[J]. EFSA Journal, 14(8):4546.

ERNY D, HRABE DE ANGELIS A L, JAITIN D, et al., 2015. Host microbiota constantly control maturation and function of microglia in the CNS[J]. Nature Neuroscience, 18(7): 965–977.

FANG C L, SUN H, WU J, et al., 2014. Effects of sodium butyrate on growth performance,haematological and immunological characteristics of weanling piglets[J]. Journal of Animal Physiology and Animal nutrition, 98(4): 680–685.

FANG S, CHEN X, YE X, et al., 2020. Effects of gut microbiome and short-chain fatty acids (SCFAs) on finishing weight of meat rabbits[J]. Frontiers in Microbiology, 11:1835.

FELIZARDO R J F, DE ALMEIDA D C, PEREIRA R L, et al., 2019. Gut microbial metabolite butyrate protects against proteinuric kidney disease through epigenetic-and GPR109a-mediated mechanisms[J]. The FASEB Journal, 33, 11894–11908.

FELIZARDO R J F, WATANABE I K M, DARDI P, et al., 2019. The interplay among gut microbiota, hypertension and kidney diseases: The role of short-chain fatty acids[J]. Pharmacological Research, 141: 366–377.

FERRAND-CALMELS M, PALHIèRE I, BROCHARD M, et al., 2014. Prediction of fatty acid profiles in cow, ewe, and goat milk by mid-infrared spectrometry[J]. Journal of Dairy Science, 97(1): 17–35.

FERREIRA T M, LEONEL A J, MELO M A, et al., 2012. Oral supplementation of butyrate reduces mucositis and intestinal permeability associated with 5-fluorouracil administration[J]. Lipids, 47(7): 669–678.

FISCHER A, SANANBENESI F, WANG X, et al., 2007. Recovery of learning and memory is associated with chromatin remodelling[J]. Nature, 447(7141): 178–182.

FLEGEL C, MANTENIOTIS S, OSTHOLD S, et al., 2013. Expression profile of ectopic olfactory receptors determined by deep sequencing[J]. PLoS ONE, 8: e55368.

FLEISCHER J, BUMBALO R, BAUTZE V, et al., 2015. Expression of odorant receptor Olfr78 in enteroendocrine cells of the colon[J]. Cell Tissue Res, 361: 697–710.

FLINT H J, DUNCAN S H, SCOTT K P, et al., 2015. Links between diet, gut microbiota composition and gut metabolism[J]. Proceedings of Nutrition Society, 74(1): 13–22.

FOX P F, LOWE T U, MCSWEENEY P L H, et al., 2015. Dairy chemistry and biochemistry[M]. Switzerland: Springer Cham Heidelberg New York Dordrecht London © Springer International.

FRAMPTON J, MURPHY K G, FROST G, et al., 2020. Short–chain fatty acids as potential regulators of skeletal muscle metabolism and function[J]. Nature Metabolism, 2(9):840–848.

FRICKER M, GOGGINS B J, MATEER S, et al., 2018. Chronic cigarette smoke exposure induces systemic hy– poxia that drives intestinal dysfunction[J]. JCI Insight, 3(3): e94040.

FROST G, SLEETH M L, SAHURI–ARISOYLU M, et al., 2014.The short–chain fatty acid acetate reduces appetite via a central homeostatic mechanism[J]. Nature Communications, 5: 3611.

GALVãO I, TAVARES L P, CORRêA R O, et al., 2018. The metabolic sensor GPR43 receptor plays a role in the control of Klebsiella pneumoniae infection in the lung[J]. Frontiers in Immunology, 9:142.

GANAPATHY V, GOPAL E, MIYAUCHI S, et al., 2005. Biological functions of SLC5A8, a candidate tumour suppressor[J]. Biochemical Society Transactions, 33(Pt 1): 237–240.

GAO W, WANG C, YU L, et al., 2019. Chlorogenic acid attenuates dextran sodium sulfate–induced ulcerative colitis in mice through MAPK/ERK/JNK pathway[J]. BioMed Research International, 2019:6769789.

GASTALDI D, BERTINO E, MONTI G, et al., 2010. Donkey’ s milk detailed lipid composition[J]. Frontiers in Bioscience, 2(2): 537–546.

GAUDIER E, RIVAL M, BUISINE M P, et al., 2009. Butyrate enemas upregulate Muc genes expression but decrease adherent mucus thickness in mice colon[J]. Physiological–Research, 58(1): 111–119.

GHOOS Y, GEYPENS B, HIELE M, et al., 1991. Analysis for short–chain carboxylic acids in feces by gas chromatography with an ion–trap detector[J]. Analytica Chimica Acta, 247(2): 223–227.

GIANLUCA P, RAFFAELE S, OLGA F, et al., 2013. High resolution 13C NMR detection of short - and medium - chain synthetic triacylglycerols used in butterfat adulteration[J]. European Journal of Lipid Science and Technology, 115: 858–864.

GOVINDARAJAN N, AGIS–BALBOA R C, WALTER J, et al., 2011. Sodium butyrate improves memory function in an Alzheimer's disease mouse model when administered at an advanced stage of disease progression[J]. Journal of Alzheimers Disease, 26(1): 187–197.

HAAK B W, LITTMANN E R, CHAUBARD J L, et al., 2018. Impact of gut colonization with butyrate–producing microbiota on respiratory viral infection following allo–HCT[J]. Blood, 131 (26):2978–2986.

HAGHIKIA A, JÖRG S, DUSCHA A, et al., 2015. Dietary fatty acids directly impact central nervous system autoimmunity via the small intestine[J]. Immunity, 43(4):817–829.

HAGHIKIA A, ZIMMERMANN F, SCHUMANN P, et al., 2021. Propionate attenuates atherosclerosis by immune-dependent regulation of intestinal cholesterol metabolism[J]. European Heart Journal, 43(6):518–533.

HENAGAN T M, STEFANSKA B, FANG Z, et al., 2015. Sodium butyrate epigenetically modulates high-fat diet-induced skeletal muscle mitochondrial adaptation, obesity and insulin resistance through nucleosome positioning[J]. British Journal of Pharmacology, 172(11):2782–2798.

HINRICHSEN F, HAMM J, WESTERMANN M, et al., 2021. Microbial regulation of hexokinase 2 links mitochondrial metabolism and cell death in colitis[J]. Cell Metabolism, 33(12): 2355–2366.

HOSOMI K, SAITO M, PARK J, et al., 2022. Oral administration of Blautia wexlerae ameliorates obesity and type 2 diabetes via metabolic remodeling of the gut microbiota[J]. Nature Communications, 13(1):4477.

HOU Y Q, LIU Y L, HU J, et al., 2006. Effects of lactitol and tributyrin on growth performance, small intestinal morphology and enzyme activity in weaned pigs[J]. Asian Australasian Journal of Animal Sciences,19(10): 1470–1477.

HOU Y, WANG L, YI D, et al., 2014. Dietary supplementation with tributyrin alleviates intestinal injury in piglets challenged with intrarectal administration of acetic acid[J]. British Journal of Nutrition, 111(10): 1748–1758.

HOYLES L, SNELLING T, UMLAI U K, et al., 2018. Microbiome-host systems interactions: protective effects of propionate upon the blood-brain barrier[J]. Microbiome, 6(1): 55.

HUANG W, GUO H L, DENG X, et al., 2017. Short-chain fatty acids inhibit oxidative stress and inflammation in mesangial cells induced by high glucose and lipopolysaccharide[J]. Experimental and Clinical Endocrinology & Diabetes, 125: 98–105.

HUUSKONEN J, SUURONEN T, NUUTINEN T, et al., 2004. Regulation of microglial inflammatory response by sodium butyrate and short-chain fatty acids[J]. British Journal of Pharmacology, 141(5): 874–880.

INDRIO F, MARTINI S, FRANCAVILLA R, et al., 2017. Epigenetic matters: The link between early nutrition, microbiome, and long-term health development[J]. Frontiers in Pediatrics, 5:178.

IWANAGA T, TAKEBE K, KATO I, et al., 2006. Cellular expression of monocarboxylate transporters (MCT) in the digestive tract of the mouse, rat, and humans, with special reference to SLC5A8[J]. Biomed Research, 27(5): 243–254.

JANG YO, KIM O H, KIM S J, et al., 2021. High-fiber diets attenuate emphysema development via modulation of gut microbiota and metabolism[J]. Scientific Reports, 11(1):7008.

JIANG X W, LI Y T, YE J Z, et al., 2020. New strain of Pediococcus pentosaceus alleviates ethanol-induced liver injury by modulating the gut microbiota and short-chain fatty acid metabolism[J]. World Journal of Gastroenterology, 26(40):6224–6240.

JIANG Z Z, LIU Y N, ZHU Y, et al., 2016. Characteristic chromatographic fingerprint study of short-chain fatty acids in human milk, infant formula, pure milk and fermented milk by gas chromatography-mass spectrometry[J]. International Journal of Food Sciences and Nutrition, 67(6): 632–640.

JIANG Z Z, LIU Y N, ZHU Y, et al., 2016. Characteristic chromatographic fingerprint study of short-chain fatty acids in human milk, infant formula, pure milk and fermented milk by gas chromatography-mass spectrometry[J]. International Journal of Food Sciences and Nutrition, 67(6): 632–640.

JIN C J, SELLMANN C, ENGSTLER A J, et al., 2015. Supplementation of sodium butyrate protects mice from the development of non–alcoholic steatohepatitis (NASH)[J]. British Journal of Nutrition, 114(11):1745–1755.

JIN L, SHI X, YANG J, et al., 2020. Gut microbes in cardiovascular diseases and their potential therapeutic applications[J]. Protein & Cell, 12(5):346–359.

JOSEPH N, VASODAVAN K, SAIPUDIN N A, et al., 2019. Gut microbiota and short–chain fatty acids (SCFAs) profiles of normal and overweight school children in Selangor after probiotics administration[J]. Journal of Functional Foods, 57:103–111.

KARNHOLZ A, KUSEL K, GOSSNER A, et al., 2002. Tolerance and metabolic response of acetogenic bacteria toward oxygen[J]. Applied and Environmental Microbiology, 68(2): 1005–1009.

KAROOR V, STRASSHEIM D, SULLIVAN T, et al., 2021. The short–chain fatty acid butyrate attenuates pulmonary vascular remodeling and inflammation in hypoxia–induced pulmonary hypertension[J]. International Journal of Molecular Sciences, 22(18):9916.

KEELY S, HANSBRO P M, 2014. Lung–gut cross talk : A potential mechanism for intestinal dysfunction in patients with COPD[J]. Chest, 145 (2):199–200.

KELLY C J, ZHENG L, CAMPBELL E L, et al., 2015. Crosstalk between microbiota–derived short–chain fatty acids and intestinal epithelial hif augments tissue barrier function[J]. Cell Host Microbe, 2015, 17(5): 662–671.

KELLY J R, MINUTO C, CRYAN J F, et al., 2017. Cross Talk: The Microbiota and Neurodevelopmental Disorders[J]. Frontiers in Neuroscience, 11:490.

KELLY J R, MINUTO C, CRYAN J F, et al., 2021. The role of the gut microbiome in the development of schizophrenia[J]. Schizophr Res, 234:4–23.

KESPOHL M, VACHHARAJANI N, LUU M, et al., 2017. The microbial metabolite butyrate induces expression of Th1–associated factors in $CD4^+$ T cells[J]. Frontiers in Immunology, 8:1036.

KHAN I T, NADEEM M, IMRAN M, et al., 2017. Antioxidant capacity and fatty acids characterization of heat treated cow and buffalo milk[J]. Lipids in Health and Disease, 16(1): 163.

KIEFFER D A, PICCOLO B D, VAZIRI N D, et al., 2016. Resistant starch alters gut microbiome and metabolomic profiles concurrent with amelioration of chronic kidney disease in rats[J]. American Journal of Physiology–Renal Physiology, 310: F857–F71.

KIMURA I, ICHIMURA A, OHUE–KITANO R, et al., 2020. Free fatty acid receptors in health and disease[J]. Physiological Reviews, 100(1): 171–210.

KOH A, DE VADDER F, KOVATCHEVA–DATCHARY P, et al., 2016. From dietary fiber to host physiology: short–chain fatty acids as key bacterial metabolites[J]. Cell, 165(6): 1332–1345.

KRATSMAN N, GETSELTER D, ELLIOTT E, 2016. Sodium butyrate attenuates social behavior deficits and modifies the transcription of inhibitory/excitatory genes in the frontal cortex of an autism model[J]. Neuropharmacology, 102: 136–145.

LANG S, SCHNABL B, 2020. Microbiota and Fatty Liver Disease–the Known, the Unknown, and the Future[J]. Cell Host & Microbe, 28(2):233–244.

LE GALL M, GALLOIS M, SEVE B, et al., 2009. Comparative effect of orally administered sodium butyrate before or after weaning on growth and several indices of gastrointestinal biology of piglets [J]. British Journal of Nutrition, 102(9): 1285–1296.

LEE H S, 2019. The interaction between gut microbiome and nutrients on development of human disease through epigenetic mechanisms[J]. Genomics and Informatics, 17(3)：e24.

LEE J, D'AIGLE J, ATADJA L, et al., 2020. Gut microbiota–derived short–chain fatty acids promote poststroke recovery in aged mice[J]. Circulation Research, 127(4):453–465.

LEWIS G, WANG B, SHAFIEI JAHANI P, et al., 2019. Dietary fiber–induced microbial short chain fatty acids suppress ILC2–dependent airway inflammation[J]. Frontiers in Immunology, 10:2051.

LEWIS K, LUTGENDORFF F, PHAN V, et al., 2010. Enhanced translocation of bacteria across metabolically stressed epithelia is reduced by butyrate[J]. Inflammatory bowel diseases, 16(7): 1138–1148.

LI B, MAO Q, ZHOU D, et al., 2021. Effects of tea against alcoholic fatty liver disease by modulating gut microbiota in chronic alcohol–exposed mice[J]. In Foods, 10(6):1232.

LI C, LIU Z, BATH C, et al., 2022. Optimised method for short–chain fatty acid profiling of bovine milk and serum[J]. Molecules, 14(2): 738–754.

LI X, FAN X, YUAN X, et al., 2021. The role of butyric acid in treatment response in drug–naïve first episode schizophrenia[J]. Front Psychiatry, 12:724664.

LI Y J, CHEN X C, KWAN T K, et al., 2020. Dietary fiber protects against diabetic nephropathy through short–chain fatty acid?Mediated activation of g protein?Coupled receptors gpr43 and gpr109a[J]. Journal of the American Society of Nephrology, 31: 1267–81.

LIU H N, REN F Z, JIANG L, et al., 2011. Short communication: Fatty acid profile of yak milk from the Qinghai–Tibetan Plateau in different seasons and for different parities[J]. Journal of Dairy Science, 94(4): 1724–1731.

LIU L, FU C, LI F, 2019. Acetate Affects the Process of Lipid Metabolism in Rabbit Liver, Skeletal Muscle and Adipose Tissue[J]. Animals, 9(10):799.

LIU Y, JIN X, HONG H G, et al., 2020. The relationship between gut microbiota and short chain fatty acids in the renal calcium oxalate stones disease[J]. Faseb Journal, 34: 11200–11214.

LIU Z, EZERNIEKS V, ROCHFORT S, et al., 2018a. Comparison of methylation methods for fatty acid analysis of milk fat[J]. Food Chemistry, 261: 210–215.

LIU Z, EZERNIEKS V, WANG J, et al., 2017. Heat stress in dairy cattle alters lipid composition of milk[J]. Scientific Reports, 7: 961.

LIU Z, ROY N C, GUO Y, et al., 2016. Human breast milk and infant formulas differentially modify the intestinal microbiota in human infants and host physiology in rats[J]. Journal of Nutrition, 146(2): 191–199.

LIU Z, WANG J, LI C, et al., 2020. Development of one–step sample preparation methods for fatty acid profiling of milk fat[J]. Food Chemistry, 315: 126281.

LOOMBA R, SANYAL A J, 2013. The global NAFLD epidemic[J]. Nature Reviews Gastroenterology & Hepatology, 10:686–690.

LOUIS P, HOLD G L, FLINT H J, 2014. The gut microbiota, bacterial metabolites and colorectal cancer[J]. Nature Reviews Microbiology, 12(10): 661–672.

LOUVET A, MATHURIN P, 2015. Alcoholic liver disease: mechanisms of injury and targeted treatment[J]. Nature Reviews Gastroenterology & Hepatology, 12(4):231–242.

LUCAS S, OMATA Y, HOFMANN J, et al., 2018. Short–chain fatty acids regulate systemic bone mass and protect from pathological bone loss[J]. Nature Communications, 9(1):55.

LYU L C, HSU C Y, YEH C Y, et al., 2003. A case–control study of the association of diet and obesity with gout in Taiwan [J]. Am J Clin Nutr, 78: 690–701.

MA X, FAN P X, LI L S, et al., 2012. Butyrate promotes the recovering of intestinal wound healing through its positive effect on the tight junctions[J]. Journal of Animal Science, 90(4): 266–268.

MACFARLANE G T, GIBSON G R, BEATTY E, et al., 1992. Estimation of short–chain fatty acid production from protein by human intestinal bacteria based on branched–chain fatty acid measurements[J]. FEMS Microbiology Ecology, 101(1992): 81–88.

MACHADO M G, SENCIO V, TROTTEIN F, 2021. Short–chain fatty acids as a potential treatment for infections: a closer look at the lungs[J]. Infection and Immunity, 89（9）: e0018821.

MACHADO R A, CONSTANTINO L D, TOMASI C D, et al., 2012. Sodium butyrate decreases the activation of nf–kappa b reducing inflammation and oxidative damage in the kidney of rats subjected to contrast–induced nephropathy[J]. Nephrology Dialysis Transplantation, 27: 3136–3140.

MADAN J C, KOESTLER D C, STANTON B A, et al., 2012. Serial analysis of the gut and respiratory microbiome in cystic fibrosis in infancy: interaction between intestinal and respiratory tracts and impact of nutritional exposures[J]. mBio, 3（4）: e00251–12.

MAO J W, TANG H Y, ZHAO T, et al., 2015. Intestinal mucosal barrier dysfunction participates in the progress of nonalcoholic fatty liver disease[J]. International journal of clinical and experimental pathology, 8(4):3648–3658.

MENG X, LI S, LI Y, et al., 2018. Gut Microbiota’s Relationship with Liver Disease and Role in Hepatoprotection by Dietary Natural Products and Probiotics[J]. In Nutrients, 10(10):1457.

MESSIAS T, ALVES S P, BESSA R J B, et al., 2021. Fatty acid profile of milk from Nordestina donkey breed raised on Caatinga pasture[J]. Journal of Dairy Research, 88(2): 205–209.

MIHAYLOVA M M, SHAW R J, 2011. The AMPK signalling pathway coordinates cell growth, autophagy and metabolism[J]. Nature Cell Biology, 13:1016–1023.

MIKAMI D, KOBAYASHI M, UWADA J, et al., 2020. Short–chain fatty acid mitigates adenine–induced chronic kidney disease via ffa2 and ffa3 pathways[J]. Biochimica Et Biophysica Acta–Molecular and Cell Biology of Lipids, 1865: 8.

MILLARD A L, MERTES P M, ITTELET D, et al., 2002. Butyrate affects differentiation, maturation and function of human monocyte-derived dendritic cells and macrophages[J]. Clinical and Experimental Immunology, 130 (2):245–255.

MITCHELL R W, ON N H, DEL BIGIO M R, et al., 2011. Fatty acid transport protein expression in human brain and potential role in fatty acid transport across human brain microvessel endothelial cells[J]. Journal of Neurochemistry, 117(4): 735–746.

MORALES F C, AMARAL M A, XAVIER I K, et al., 2021. Short chain fatty acids (SCFAs) improves TNBS-induced colitis in zebrafish[J]. Current Research Immunology, 2: 142–154.

MOUZAKI M, COMELLI E M, ARENDT B M, et al., 2013. Intestinal microbiota in patients with nonalcoholic fatty liver disease[J]. Hepatology, 9(23): 6654–6662.

MUN S J, LEE J, CHUNG K S, et al., 2021. Effect of microbial short-chain fatty acids on CYP3A4-mediated metabolic activation of human pluripotent stem cell-derived liver organoids[J]. Cells, 10(1):126.

MURASHIGE D, JANG C, NEINAST M, et al., 2020. Comprehensive quantification of fuel use by the failing and nonfailing human heart[J]. Science, 370(6514):364–368.

MURUGESAN S, NIRMALKAR K, HOYO-VADILLO C, et al., 2017. Gut microbiome production of short-chain fatty acids and obesity in children[J]. European Journal of Clinical Microbiology & Infectious Diseases, 37(4):621–625.

NATARAJAN N, HORI D, FLAVAHAN S, et al., 2016. Microbial short chain fatty acid metabolites lower blood pressure via endothelial G protein-coupled receptor 41[J]. Physiological Genomics, 48(11):826–834.

NIE P, PAN B, AHMAD M J, et al., 2022. Summer buffalo milk produced in China: A desirable diet enriched in polyunsaturated fatty acids and amino acids[J]. Foods, 11(21):3475.

OBATA Y, FURUSAWA Y, HASE K, 2015. Epigenetic modifications of the immune system in health and disease[J]. Immunology & Cell Biology, 93(3): 226–232.

OHARA T M T, 2019. Antiproliferative effects of short-chain fatty acids on human colorectal cancer cells via gene expression inhibition[J]. Anticancer Research, 39(9):4659–4666.

OLDENDORF W H, 1973. Carrier-mediated blood-brain barrier transport of short-chain monocarboxylic organic acids[J]. American Journal of Physiology, 224(6): 1450–1453.

ONYSZKIEWICZ M, GAWRYS-KOPCZYNSKA M, KONOPELSKI P, et al., 2019. Butyric acid, a gut bacteria metabolite, lowers arterial blood pressure via colon-vagus nerve signaling and GPR41/43 receptors[J]. Pflügers Archiv – European Journal of Physiology, 471(11–12):1441–1453.

ONYSZKIEWICZ M, GAWRYS-KOPCZYNSKA M, SAŁAGAJ M, et al., 2020. Valeric acid lowers arterial blood pressure in rats[J]. European Journal of Pharmacology, 877:173086.

PAIVA I, PINHO R, PAVLOU M A, et al., 2017. Sodium butyrate rescues dopaminergic cells from alpha-synuclein-induced transcriptional deregulation and DNA damage[J]. Human Molecular Genetics, 26(12): 2231–2246.

PALM C L, NIJHOLT K T, BAKKER B M, et al., 2022. Short-Chain Fatty Acids in the Metabolism of Heart Failure – Rethinking the Fat Stigma[J]. Frontiers in Cardiovascular Medicine, 9:915102.

PARADA VENEGAS D, DE LA FUENTE M K, LANDSKRON G, et al., 2019. Short Chain Fatty Acids (SCFAs)-Mediated Gut Epithelial and Immune Regulation and Its Relevance for Inflammatory Bowel Diseases[J]. Frontiers in Immunology, 10:1486.

PARK J, GOERGEN C J, HOGENESCH H, et al., 2016. Chronically elevated levels of short-chain fatty acids induce t cell-mediated ureteritis and hydronephrosis[J]. Journal of Immunology, 196: 2388-2400.

PATNALA R, ARUMUGAM T V, GUPTA N, et al., 2017. HDAC Inhibitor Sodium Butyrate-Mediated Epigenetic Regulation Enhances Neuroprotective Function of Microglia During Ischemic Stroke[J]. Molecular Neurobiology, 54(8): 6391-6411.

PATTERSON A M, MULDER I E, TRAVIS A J, et al., 2017. Human Gut Symbiont Roseburia hominis Promotes and Regulates Innate Immunity[J]. Frontiers in Immunology, 8:1166.

PEGOLO S, STOCCO G, MELE M, et al., 2017. Factors affecting variations in the detailed fatty acid profile of Mediterranean buffalo milk determined by 2-dimensional gas chromatography[J]. Journal of Dairy Science, 100(4): 2564-2576.

PELASEYED T, BERGSTROM J H, GUSTAFSSON J K, et al., 2014. The mucus and mucins of the goblet cells and enterocytes provide the first defense line of the gastrointestinal tract and interact with the immune system[J]. Immunological Reviews, 260(1): 8-20.

PETRACHE I, PETRUSCA D N, 2013. The involvement of sphingolipids in chronic obstructive pulmonary diseases[J]. Metabolic Control (216):247-264.

PIETRZAK-FIÉCKO R, KAMELSKA-SADOWSKA A M, 2020. The comparison of nutritional value of human milk with other mammals' milk[J]. Nutrients, 12(5): 1404.

PLUZNICK J L, PROTZKO R J, GEVORGYAN H, et al., 2013. Olfactory receptor responding to gut microbiota-derived signals plays a role in renin secretion and blood pressure regulation[J]. Proc. Natl.Acad. Sci. USA, 110: 4410–4415.

POLL B G, XU J, JUN S, et al., 2021. Acetate, a short-chain fatty acid, acutely lowers heart rate and cardiac contractility along with blood pressure[J]. Journal of Pharmacology and Experimental Therapeutics, 377(1): 39-50.

POLYVIOU T, MACDOUGALL K, CHAMBERS ES, et al., 2016. Randomised clinical study: inulin short-chain fatty acid esters for targeted delivery of short-chain fatty acids to the human colon[J]. Aliment Pharmacol Ther, 44(7):662-672.

PRENTICE P M, SCHOEMAKER M H, VERVOORT J, et al., 2019. Human milk short-chain fatty acid composition is associated with adiposity outcomes in infants[J]. Journal of Nutrition, 149(5): 716-722.

PTACEK M, MILERSKI M, DUCHACEK J, et al., 2019. Analysis of fatty acid profile in milk fat of Wallachian sheep during lactation[J]. Journal of Dairy Research, 86(2): 233-237.

RAHMAN M M, KUKITA A, KUKITA T, et al., 2003. Two histone deacetylase inhibitors, trichostatin A and sodium butyrate, suppress differentiation into osteoclasts but not into macrophages[J]. Blood, 101(9):3451–3459.

RATAJCZAK W, RYŁ A, MIZERSKI A, et al., 2019. Immunomodulatory potential of gut microbiome–derived shortchain fatty acids (SCFAs)[J]. Acta Biochimica Polonica, 66 (1):1–12.

RATERINK R J, LINDENBURG P W, VREEKEN R J, et al., 2014. Recent developments in sample–pretreatment techniques for mass spectrometry–based metabolomics[J]. Trends in Analytical Chemistry, 61: 157–167.

REICHARDT N, DUNCAN S H, YOUNG P, et al., 2014. Phylogenetic distribution of three pathways for propionate production within the human gut microbiota[J]. ISME Journal, 8(6): 1323–1335.

REN M, LI H, FU Z, et al., 2022. Centenarian–sourced Lactobacillus casei combined with dietary fiber complex ameliorates brain and gut function in aged mice[J]. Nutrients, 14(2):324.

RESENDE W R, VALVASSORI S S, REUS G Z, et al., 2013. Effects of sodium butyrate in animal models of mania and depression: implications as a new mood stabilizer[J]. Behavioural Pharmacology, 2013, 24(7): 569–579.

REY F E, FAITH J J, BAIN J, et al., 2010. Dissecting the *in vivo* metabolic potential of two human gut acetogens[J]. Journal of Biological Chemistry, 285(29): 22082–22090.

RICHARDS L B, LI M, FOLKERTS G, et al., 2020. Butyrate and propionate restore the cytokine and house dust mite compromised barrier function of human bronchial airway epithelial cells[J]. International Journal of Molecular Sciences, 22(1):65.

RICKE S. 2003. Perspectives on the use of organic acids and short chain fatty acids as antimicrobials[J]. Poultry Science 82 (4):632–639.

RíOS–COVIáN D, RUAS–MADIEDO P, MARGOLLES A, et al., 2016. Intestinal short chain fatty acids and their link with diet and human health[J]. Frontiers in Microbiology, 7:185.

ROOKS M G, GARRETT W S, 2016. Gut microbiota, metabolites and host immunity[J]. Nature Reviews Immunology, 16 (6):341–352.

RUTTING S, XENAKI D, MALOUF M, et al., 2018. Short–chain fatty acids increase TNF α –induced inflammation in primary human lung mesenchymal cells through the activation of p38 MAPK[J]. American Journal of Physiology–Lung Cellular and Molecular Physiology, 316(1): L157–L174.

SADLER R, CRAMER J V, HEINDL S, et al., 2020. Short–chain fatty acids improve poststroke recovery via immunological mechanisms[J]. Journal of Neuroscienc, 40(5): 1162–1173.

SAHURI–ARISOYLU M, BRODY L P, PARKINSON J R, et al., 2016. Reprogramming of hepatic fat accumulation and 'browning' of adipose tissue by the short–chain fatty acid acetate[J]. International Journal of Obesity, 40(6):955–963.

SAJDEL–SULKOWSKA E M, 2021. Neuropsychiatric ramifications of COVID–19:Short–chain fatty acid deficiency and disturbance of microbiota–gut–brain axis signaling[J]. BioMed Research International, 2021:

1–15.

SCHONFELD P, WOJTCZAK L, 2016. Short– and medium–chain fatty acids in energy metabolism: the cellular perspective [J]. Journal of Lipid Research, 57(6): 943–954.

SCHROEDER F A, LIN C L, CRUSIO W E, et al., 2007. Antidepressant–like effects of the histone deacetylase inhibitor, sodium butyrate, in the mouse[J]. Biological Psychiatry, 62(1): 55–64.

SEHRAWAT T S, LIU M, SHAH V H, 2020. The knowns and unknowns of treatment for alcoholic hepatitis[J]. The Lancet Gastroenterology & Hepatology, 5(5):494–506.

SENCIO V, BARTHELEMY A, TAVARES L P, et al., 2020. Gut Dysbiosis during Influenza Contributes to Pulmonary Pneumococcal Superinfection through Altered Short–Chain Fatty Acid Production[J]. Cell Reports, 30 (9):2934–2947.

SHAIDULLOV I F, SOROKINA D M, SITDIKOV F G, et al., 2021. Short chain fatty acids and colon motility in a mouse model of irritable bowel syndrome[J]. BMC Gastroenterology, 21(1): 37–48.

SHARMA S, TALIYAN R, SINGH S, 2015. Beneficial effects of sodium butyrate in 6–OHDA induced neurotoxicity and behavioral abnormalities: modulation of histone deacetylase activity[J]. Behavioural Brain Research, 291: 306–314.

SHARMA S, TALIYAN R, SINGH S, 2015. Beneficial effects of sodium butyrate in 6–OHDA induced neurotoxicity and behavioral abnormalities: Modulation of histone deacetylase activity[J]. Behavioural Brain Research, 291:306–314.

SHI H, GE X, MA X, et al., 2021. A fiber–deprived diet causes cognitive impairment and hippocampal microglia–mediated synaptic loss through the gut microbiota and metabolites[J]. Microbiome, 9(1): 223.

SILVA Y P, BERNARDI A, FROZZA R L, 2020. The role of short–chain fatty acids from gut microbiota in gut–brain communication[J]. Frontiers in Endocrinology(Lausanne), 11:25.

SINGH N, GURAV A, SIVAPRAKASAM S, et al., 2014. Activation of Gpr109a, receptor for niacin and the commensal metabolite butyrate, suppresses colonic inflammation and carcinogenesis[J]. Immunity, 40 (1):128–139.

SIVAPRAKASAM S, BHUTIA Y D, YANG S, et al., 2017. Short–chain fatty acid transporters: role in colonic homeostasis[J]. Comprehensive Physiology, 8(1): 299–314.

SMITH P M, HOWITT M R, PANIKOV N, et al., 2013. The microbial metabolites, short–chain fatty acids, regulate colonic T reg cell homeostasis[J]. Science, 341 (6145):569–573.

SOLIMAN M L, PUIG K L, COMBS C K, et al., 2012. Acetate reduces microglia inflammatory signaling *in vitro*[J]. Journal of Neurochemistry, 123(4): 555–567.

SOLIMAN M L, ROSENBERGER T A, 2011. Acetate supplementation increases brain histone acetylation and inhibits histone deacetylase activity and expression [J]. Molecular and Cellular Biochemistry, 352(1/2): 173–180.

STAFFORD J M, RAYBUCK J D, RYABININ A E, et al.,2012. Increasing histone acetylation in the hippocampus–infralimbic network enhances fear extinction[J]. Biological Psychiatry, 72(1): 25–33.

STINSON L F, GAY M C L, KOLEVA P T, et al., 2020. Human milk from atopic mothers has lower levels of short chain fatty acids[J]. Frontiers in Immunology, 11: 1427.

STINSON L F, GAY M C L, KOLEVA P T, et al., 2020. Human milk from atopic mothers has lower levels of short chain fatty acids[J]. Frontiers in Immunology, 11: 1427.

STUMPFF F, 2018. A look at the smelly side of physiology: transport of short chain fatty acids[J]. Pflugers Archiv–European Journal of Physiology, 470(4): 571–598.

SUN X F, ZHANG B M, HONG X, et al., 2013. Histone deacetylase inhibitor, sodium butyrate, attenuates gentamicin–induced nephrotoxicity by increasing prohibitin protein expression in rats[J]. European Journal of Pharmacology, 707: 147–54.

SUN Y Y, ZHOU C X, CHEN Y M, et al., 2022. Quantitative increase in short–chain fatty acids, especially butyrate protects kidney from ischemia/reperfusion injury[J]. Journal of Investigative Medicine, 70: 29–35.

TAMANG J P, SHIN D H, JUNG S J, et al., 2016. Functional properties of microorganisms in fermented foods[J]. Frontiers in Microbiology, 7:578.

TAN J K, MACIA L, MACKAY C R, 2023. Dietary fiber and SCFAs in the regulation of mucosal immunity[J]. J Allergy Clin Immunol, 151(2):361–370.

TANG G, DU Y, GUAN H, et al., 2021. Butyrate ameliorates skeletal muscle atrophy in diabetic nephropathy by enhancing gut barrier function and FFA2–mediated PI3K/Akt/mTOR signals[J]. British Journal of Pharmacology, 179(1):159–178.

TAZOE H, OTOMO Y, KARAKI S I, et al., 2009. Expression of short–chain fatty acid receptor GPR41 in the human colon[J]. Biomedical Research, 30(3):149–156.

TEIXEIRA M I, ANDRADE L R, FARINA M, et al., 2004. Characterization of short chain fatty acid microcapsules produced by spray drying[J]. Materials Science and Engineering: C, 24: 653–658.

TENG F, WANG P, YANG L, et al., 2017. Quantification of fatty acids in human, cow, buffalo, goat, yak, and camel milk using an improved one–step GC–FID method[J]. Food Analytical Methods, 10(8): 2881–2891.

THORBURN A N, HANSBRO P M, 2010. Harnessing regulatory T cells to suppress asthma: From Potential to therapy[J]. American Journal of Respiratory Cell and Molecular Biology, 43 (5):511–519.

TROMPETTE A, GOLLWITZER E S, PATTARONI C, et al., 2018. Dietary Fiber Confers Protection against Flu by Shaping Ly6c Patrolling Monocyte Hematopoiesis and CD8 T Cell Metabolism[J]. Immunity, 48(5):992–1005.

TROMPETTE A, GOLLWITZER E S, YADAVA K, et al., 2014. Gut microbiota metabolism of dietary fiber influences allergic airway disease and hematopoiesis[J]. Nature Medicine, 20 (2):159–166.

VAN DE WOUW M, BOEHME M, LYTE J M, et al., 2018. Short–chain fatty acids: microbial metabolites that alleviate stress-induced brain-gut axis alterations[J].The Journal of Physiology, 596(20): 4923-4944.

VAN IMMERSEEL F, DE BUCK J, PASMANS F, et al., 2003. Invasion of *Salmonella enteritidis* in avian intestinal epithelial cells in vitro is influenced by short-chain fatty acids[J]. International Journal of Food Microbiology, 85 (3):237-248.

VIEIRA A T, GALVÃO I, MACIA L M, et al., 2017. Dietary fiber and the short-chain fatty acid acetate promote resolution of neutrophilic inflammation in a model of gout in mice [J]. J Leukoc Biol, 101: 275-284.

WAGHULDE H, CHENG X, GALLA S, et al., 2018. Attenuation of Microbiotal Dysbiosis and Hypertension in a CRISPR/Cas9 Gene Ablation Rat Model of GPER1[J]. Hypertension, 72(5):1125-1132.

WAN Z X, WANG X L, XU L, et al., 2010. Lipid content and fatty acids composition of mature human milk in rural North China[J]. British Journal of Nutrition, 103(6): 913-916.

WANG F G, CHEN M Q, LUO R B, et al., 2022a. Fatty acid profiles of milk from Holstein cows, Jersey cows, buffalos, yaks, humans, goats, camels, and donkeys based on gas chromatography-mass spectrometry[J]. Journal of Dairy Science, 105(2): 1687-1700.

WANG L, LI X, HUSSAIN M, et al., 2020. Effect of lactation stages and dietary intake on the fatty acid composition of human milk (A study in northeast China)[J]. International Dairy Journal, 101: 104580.

WANG P, ZHANG Y, GONG Y, et al., 2018. Sodium butyrate triggers a functional elongation of microglial process via Akt-small RhoGTPase activation and HDACs inhibition[J]. Neurobiology of Disease, 111: 12-25.

WANG R X, LEE J S, CAMPBELL E L, et al., 2020. Microbiota-derived butyrate dynamically regulates intestinal homeostasis through regulation of actin-associated protein synaptopodin[J]. Proceedings of the National Academy of Sciences of the United States of America, 117(21): 11648-11657.

WANG S Q, LV D, JIANG S H, et al., 2019. Quantitative reduction in short-chain fatty acids, especially butyrate, contributes to the progression of chronic kidney disease[J]. Clinical Science, 133: 1857-70.

WELLS J M, BRUMMER R J, DERRIEN M, et al., 2017. Homeostasis of the gut barrier and potential biomarkers[J]. American Journal Physiology Gastrointestinal Liver Physiology, 312(3): 171-193.

WEN Z S L J, ZOU X T, 2012. Effects of sodium butyrate on the intestinal morphology and DNA-binding activity of intestinal nuclear factor-κB in weanling pigs[J]. Asian Journal of Animal and Veterinary Advances, 11: 814-821.

WIKING L, LØKKE M M, KIDMOSE U, et al., 2017. Comparison between novel and standard methods for analysis of free fatty acids in milk - Including relation to rancid flavour[J]. International Dairy Journal, 75: 22-29.

WONG R J, CHEUNG R, AHMED A, 2014. Nonalcoholic steatohepatitis is the most rapidly growing indication for liver transplantation in patients with hepatocellular carcinoma in the U.S[J]. Hepatology, 100(3):607-612.

XU Q, ZHANG R, MU Y, et al., 2022. Propionate ameliorates alcohol-induced liver injury in mice via the gut-liver axis: Focus on the improvement of intestinal permeability[J]. Journal of Agricultural and Food Chemistry,

70(20):6084–6096.

XU Y, ZHU Y, LI X, et al., 2020. Dynamic balancing of intestinal short–chain fatty acids: the crucial role of bacterial metabolism[J]. Trends in Food Science & Technology, 2020, 100: 118–130.

YAMAWAKI Y, YOSHIOKA N, NOZAKI K, et al., 2018. Sodium butyrate abolishes lipopolysaccharide–induced depression–like behaviors and hippocampal microglial activation in mice[J]. Brain Research, 1680: 13–38.

YANG F, CHEN H, GAO Y, et al., 2020. Gut microbiota–derived short–chain fatty acids and hypertension: Mechanism and treatment[J]. Biomedicine & Pharmacotherapy, 130:110503.

YANG L L, MILLISCHER V, RODIN S, et al., 2020. Enteric short–chain fatty acids promote proliferation of human neural progenitor cells[J]. Journal of Neurochemistry, 154(6):635–646.

YAO J, CHEN Y, XU M, 2022. The critical role of short–chain fatty acids in health and disease: A subtle focus on cardiovascular disease–NLRP3 inflammasome–angiogenesis axis[J]. Clinical Immunology, 238:109013.

YE J, LV L, WU W, et al., 2018. Butyrate protects mice against methionine–choline–deficient diet–induced non–alcoholic steatohepatitis by improving gut barrier function, attenuating inflammation and reducing endotoxin levels[J]. Frontiers in Microbiology, 9:1967.

YOSHIDA H, ISHII M, AKAGAWA M, 2019. Propionate suppresses hepatic gluconeogenesis via GPR43/AMPK signaling pathway[J]. Archives of Biochemistry and Biophysics, 672:108057.

YOSHIKAWA S, ARAOKA R, KAJIHARA Y, et al., 2018. Valerate production by *Megasphaera elsdenii* isolated from pig feces[J]. J Biosci Bioeng, 125(5): 519–524.

YU L, ZHONG X, HE Y, et al., 2020. Butyrate, but not propionate, reverses maternal diet–induced neurocognitive deficits in offspring[J]. Pharmacological Research, 160: 105082.

ZAKY A, GLASTRAS S J, WONG M Y W, et al., 2021. The role of the gut microbiome in diabetes and obesity–related kidney disease[J]. International journal of molecular sciences, 22(17): 9641.

ZEBARI H M, RUTTER S M, BLEACH E C L, 2019. Fatty acid profile of milk for determining reproductive status in lactating Holstein Friesian cows[J]. Animal Reproduction Science, 202: 26–34.

ZENG X, SUNKARA L T, JIANG W, et al., 2013. Induction of porcine host defense peptide gene expression by short–chain fatty acids and their analogs[J]. PLoS One, 8(8): e72922.

ZHANG H, QIN S, ZHU Y, et al., 2022. Dietary resistant starch from potato regulates bone mass by modulating gut microbiota and concomitant short–chain fatty acids production in meat ducks[J]. Frontiers in Nutrition, 9: 860086.

ZHANG J, CHENG S, WANG Y, et al., 2013. Identification and characterization of the free fatty acid receptor 2 (FFA2) and a novel functional FFA2–like receptor (FFA2L) for short–chain fatty acids in pigs: Evidence for the existence of a duplicated FFA2 gene (FFA2L) in some mammalian species[J]. Domest. Anim. Endocrinol, 47: 108–118.

ZHANG Y, LI J X, ZHANG Y, et al., 2021. Intestinal microbiota participates in nonalcoholic fatty liver disease progression by affecting intestinal homeostasis[J]. World J Clin Cases, 9(23): 6654–6662.

ZHAO S T, WANG C Z, 2018. Regulatory T cells and asthma[J]. Journal of Zhejiang University: Science B, 19 (9):663–673.

ZHONG C Y, DAI Z W, CHAI L X, et al., 2021. The change of gut microbiota–derived short–chain fatty acids in diabetic kidney disease[J]. Journal of Clinical Laboratory Analysis , 35: 11.

ZHOU D, PAN Q, XIN F Z, et al., 2017. Sodium butyrate attenuates high–fat diet–induced steatohepatitis in mice by improving gut microbiota and gastrointestinal barrier[J]. World Journal of Gastroenterology, 23(1):60–75.

ZHOU H, SUN J, GE L, et al., 2020. Exogenous infusion of short–chain fatty acids can improve intestinal functions independently of the gut microbiota[J]. Journal of Animal Science, 98(12): 1–10.

ZHU F, GUO R, WANG W, et al., 2020. Transplantation of microbiota from drug–free patients with schizophrenia causes schizophrenia–like abnormal behaviors and dysregulated kynurenine metabolism in mice[J]. Mol Psychiatry, 25(11):2905–2918.

ZHU X, TAO Y, LIANG C, et al., 2015. The synthesis of n–caproate from lactate: a new efficient process for medium–chain carboxylates production[J]. Scientific Reports, 5: 14360.

ZUMBRUN S D, MELTON–CELSA A R, O'BRIEN A D, 2014. When a healthy diet turns deadly[J]. Gut Microbes, 5: 40–3.

ZUMBRUN S D, MELTON–CELSA A R, SMITH M A, et al., 2013. Dietary choice affects shiga toxin–producing *Escherichia coli* (stec) O157:H7 colonization and disease[J]. Proceedings of the National Academy of Sciences of the United States of America, 110: E2126–E2133.